Lecture Notes in Civil Engineering

Volume 204

Lecture Notes in Civil Engineering (LNCE) publishes the latest developments in Civil Engineering—quickly, informally and in top quality. Though original research reported in proceedings and post-proceedings represents the core of LNCE, edited volumes of exceptionally high quality and interest may also be considered for publication. Volumes published in LNCE embrace all aspects and subfields of, as well as new challenges in, Civil Engineering. Topics in the series include:

- Construction and Structural Mechanics
- Building Materials
- Concrete, Steel and Timber Structures
- Geotechnical Engineering
- Earthquake Engineering
- Coastal Engineering
- Ocean and Offshore Engineering; Ships and Floating Structures
- Hydraulics, Hydrology and Water Resources Engineering
- Environmental Engineering and Sustainability
- Structural Health and Monitoring
- Surveying and Geographical Information Systems
- Indoor Environments
- Transportation and Traffic
- Risk Analysis
- Safety and Security

To submit a proposal or request further information, please contact the appropriate Springer Editor:

– Pierpaolo Riva at pierpaolo.riva@springer.com (Europe and Americas);
– Swati Meherishi at swati.meherishi@springer.com (Asia - except China, and Australia, New Zealand);
– Wayne Hu at wayne.hu@springer.com (China).

All books in the series now indexed by Scopus and EI Compendex database!

More information about this series at https://link.springer.com/bookseries/15087

Magd Abdel Wahab
Editor

Proceedings of the 2nd International Conference on Structural Damage Modelling and Assessment

SDMA 2021, 4–5 August, Ghent University, Belgium

Springer

Editor
Magd Abdel Wahab
Faculty of Engineering and Architecture
Ghent University, Laboratory Soete
Zwijnaarde, Belgium

ISSN 2366-2557 ISSN 2366-2565 (electronic)
Lecture Notes in Civil Engineering
ISBN 978-981-16-7215-6 ISBN 978-981-16-7216-3 (eBook)
https://doi.org/10.1007/978-981-16-7216-3

This Springer imprint is published by the registered company Springer Nature Singapore Pte Ltd.
The registered company address is: 152 Beach Road, #21-01/04 Gateway East, Singapore 189721, Singapore

Organising Committee

Chairman

Prof. Magd Abdel Wahab, Ghent University, Belgium

International Scientific Committee

Prof. S. Abdullah, Universiti Kebangsaan Malaysia, Malaysia
Dr. H. T. Ali, University of Bristol, UK
Dr. I. Hilmy, International Islamic University Malaysia
Prof. G-R. Gillich, Eftimie Murgu Univ Resita, Romania
Dr. S. Khatir, Ghent University, Belgium
Dr. C. Le Thanh, Open University Ho Chi Minh City, Vietnam
Prof. N-A. Noda, Kyushu Institute of Technology, Japan
Prof. K. Oda, Oita University, Japan
Prof. R. V. Prakash, Indian Institute of Technology, India
Prof. T. Rabczuk, Bauhaus University Weimar, Germany
Prof. A. Rudawska, Lublin University of Technology, Poland
Prof. J. Toribio, University of Salamanca, Spain
Dr. L. V. Tran, Sejong University, South Korea
Prof. L. Vanegas Useche, Universidad Tecnológica de Pereira, Colombia
Dr. C. Wang, Liaocheng University, China
Prof. H-N. Xuan, Hutech University, Vietnam
Dr. X. Zhuang, Leibniz Unversität Hannover, Germany
Prof. Yongtao Bai, Chongqing University, China
Prof. Maria Korovina, Lomonosov Moscow State University, Russia
Prof. Dr. Hovik A. Matevossian, Russian Academy of Sciences, Russia
Dr. Y. L. Zhou to Xi'an Jiaotong University, China

Preface

This volume contains the proceedings of the 2nd International Conference on Structural Damage Modelling and Assessment (SDMA 2021), August 4–5, 2021, Online and at Ghent University, Belgium. The conference is a major international forum for research topics relevant to damage modelling and assessment of engineering structures and systems including numerical simulations, signal processing of sensor measurements and theoretical techniques, as well as, experimental case studies. The presentations of SDMA 2021 are divided into 2 main sessions, namely: (1) Damage in Civil Engineering and (2) Damage in Mechanical and Materials Engineering.

The organising committee is grateful to keynote speaker: Professor Filippo Berto, Department of mechanical and industrial engineering, NTNU, Trondheim Norway, for his presentation entitled 'Plant-cell-inspired interlocking structures: recent developments and future outcomes'.

Special thanks go to members of the Scientific Committee of SDMA 2021 for reviewing the articles published in this volume and for judging their scientific merits. Based on the comments of reviewers and the scientific merits of the submitted manuscripts, the articles were accepted for publication in the conference proceedings and for presentation at the conference venue. The accepted papers are of a very high scientific quality and contribute to advancement of knowledge in all research topics relevant to SDMA conference.

Finally, the organising committee would like to thank all authors, who have contributed to this volume and presented their research work at SDMA 2021.

Zwijnaarde, Belgium

Magd Abdel Wahab
Chairman of SDMA 2021

Contents

Damage in Civil Engineering

Finite Element Model Updating of Lifeline Truss Bridge Using Vibration-Based Measurement Data and Balancing Composite Motion Optimization

Lan Ngoc-Nguyen, Samir Khatir, Hoa Ngoc-Tran, Hieu Nguyen-Tran, Binh Duc-Nguyen, Thanh Bui-Tien, and Magd Abdel Wahab

Abstract Located in the heart of Hanoi (Vietnam), Chuong Duong bridge is a major truss bridge that connects one of the most heavily trafficked routes in the country-the 1A National road. Being built in the 80s of the twentieth century, after nearly 40 years of service, degradation and damages have threatened the integrity and safety of the structure. Physical and numerical evaluation of the bridge is required for the maintenance process. In this paper, we proposed a new approach to model updating of Chuong Duong bridge using vibration-based measurement data and Balancing Composite Motion Optimization (BCMO). BCMO is a newly developed meta-heuristic optimization algorithm based on individual's balancing composite motion properties which has proved to provide highly-accurate result in determining the optimal solution in mathematical problem. BCMO is applied to update the different parameters of the baseline numerical model of Chuong Duong bridge, followed by comparing the obtained dynamic properties of the updated bridge with the measured one. The final result shows that BCMO has comprehensively updated the model with a high level of accuracy, thus could be potential used to solve practical problems of lifeline structures.

Keywords Model updating · Balancing composite motion optimization · Truss bridge · Structural health monitoring

L. Ngoc-Nguyen (✉) · S. Khatir · H. Ngoc-Tran · H. Nguyen-Tran · M. Abdel Wahab
Soete Laboratory, Faculty of Engineering and Architecture, Ghent University, Ghent, Belgium
e-mail: ngoclan.nguyen@ugent.be

L. Ngoc-Nguyen · H. Ngoc-Tran · B. Duc-Nguyen · T. Bui-Tien
Faculty of Civil Engineering, University of Transport and Communications, Hanoi, Vietnam

H. Nguyen-Tran
Faculty of Information Technology, University of Transport and Communications, Hanoi, Vietnam

M. Abdel Wahab (ed.), *Proceedings of the 2nd International Conference on Structural Damage Modelling and Assessment*, Lecture Notes in Civil Engineering 204,
https://doi.org/10.1007/978-981-16-7216-3_1

1 Introduction

Recently, vibration-based structural health monitoring (VBSHM) for civil structures has been researched and applied continuously, especially to lifeline infrastructures which are vulnerable to environmental damages such as floods, typhoon, etc. The multiple types of sensors installed on the structures during monitoring and measurement not only help to evaluate accurately the dynamic behavior of the structure but also enable engineers to predict possible damages which may threaten the integrity of the structure. To ensure the effectiveness of SHM system, it is vital to have a benchmark finite-element (FE) model of the structure based on the dynamic properties of the real-life structure such as frequencies, mode shapes or damping ratio. However, the correlation between the FE model and the physical structures are always hindered by different uncertainties from both real-life conditions and simulated programs. Researchers have successfully applied model updating to tackle this problem. Teughels and De Roeck [1] used Young Modulus and shear modulus as updating parameters for Z24 bridge. The results show an improvement in accuracy of 4% and 7% for undamaged a damaged model from the initial models. Jang et al. [2] updated the corrosion condition of the historic bridge with SQP algorithm. The proposed method helps to estimate correctly the corrosion level of the test bridge. Currently, with the help of advanced optimization algorithms [3–8], researchers have been able to enhance the accuracy of model updating. Tiachacht et al. [9] applied slime mould algorithm to solve damage detection of beam and planar truss bar. The result shows that the proposed algorithm helps to detect accurately location and level of damages. Tran et al. [10] applied PSO and GA to update unknown parameters of Nam O bridge. The result shows that PSO is more superior in both accuracy level and computational cost. Ho et al. [11] combined a proposed hybrid algorithm with Morris method to update FE model of a cable-stayed bridge with high accuracy and reduced calculation time. Nguyen et al. [12] further applied GA beyond model updating of truss bridge and combine with ANN to detect damage on the structure successfully.

In this paper, we proposed an application of a recently developed optimization algorithm- Balancing Composite Motion Optimization [13]. BCMO is a newly developed meta-heuristic optimization algorithm based on individual's balancing composite motion properties which has proved to provide highly-accurate result in determining the optimal solution in mathematical problem. For our case, vibration-based measurement data of Chuong Duong bridge will be combine with this algorithm for model updating of the FE model of the bridge. The proposed method will be compared with the other optimization algorithms for effectiveness and accuracy.

2 Balancing Composite Motion Optimization

BCMO is developed based on the assumption that the optimal solution can be found in a Cartesian space while the searching movement of candidate solutions are the same in global search as well as local search. These assumptions enable the best candidate solutions per generation to either deepen its search within the local space or move to another space to continue the search. A probabilistic mathematical model is created to manage the movement of each individuals. Once each individual achieves balancing between its searching and exploiting movement, the searching ability of the whole population can also be balanced, thus optimal solution can be obtained.

BCMO is expressed in the following three steps:

- Step 1: Initialization.

Distribution of population is calculated according to the following equation:

$$x_i = x_i^L + \text{rand}(1,\, d) \times \left(x_i^U - x_i^L\right) \tag{1}$$

where x_i^L, x_i^U are the lower and upper boundary of the ith individual and d is the number of input parameters.

- Step 2: Determination of instant global point and best individual.

In this step, the movement vector of the *i*th individual of each generation with respect to global optimization point O, hereby denoted as v_i can be calculated as follow:

$$v_i = v_{i/j} + v_j \tag{2}$$

where v_j is the movement vector of the *j*th individual of each generation with respect to O, $v_{i/j}$ is the relative movement vector of the *i*th individual with respect to the jth one.

Equation (2) is validated only once the location of O is defined. Hence to overcome this difficulty, instant global point O_{in} is introduced and can be obtained by the Eq. (3) below:

$$\mathbf{x}_{\text{Oin}}^t = \begin{cases} \text{u}_1^t \text{ if } f\left(\text{u}_1^t\right) < f\left(\text{x}_1^{t-1}\right) \\ \text{x}_1^{t-1} \text{ otherwise} \end{cases} \tag{3}$$

where the best individual of current generation u_1^t is calculated based on the population information of previous generation:

$$\text{u}_1^t = \frac{LB + UB}{2} + \text{v}_{k1/k2}^t + \text{v}_{k2/1}^t \tag{4}$$

With LB, UB are lower and upper boundary of the search space respectively.

- Step 3: Calculation of composite motion of individual in solution space.

  In step 3, the movement of global search $\mathbf{v}_j$ is determined as follow

$$\mathbf{v}_j = \boldsymbol{\alpha}_j\left(\mathbf{x}_{\text{Oin}} - \mathbf{x}_j\right) \tag{5}$$

where $\boldsymbol{\alpha}_j$ is the first order derivative of distance between O_{in} and jth individual as:

$$\boldsymbol{\alpha}_j = L_{GS} \times \mathbf{dv}_j \tag{6}$$

L_{GS} = global step size scaling the movement of the jth individual.
$\mathbf{dv}_j$ = direction vector, sign of which is positive or negative with a probability of 0.5.

Similarly, to Eqs. (5) and (6), we can calculate the relative movement of the *i*th individual with respect to the *j*th one by:

$$\mathbf{v}_{i/j} = \boldsymbol{\alpha}_{ij}\left(\mathbf{x}_j - \mathbf{x}_i\right) \tag{7}$$

Finally; the position of *i*th individual in the next generation is obtained in the Eq. (8) below:

$$\mathbf{x}_i^{t+1} = \mathbf{x}_i^t + \mathbf{v}_{i/j} + \mathbf{v}_j \tag{8}$$

3 Model Updating of Chuong Duong Bridge Using Measurement Data and BCMO

3.1 Measurement Campaign of Chuong Duong Bridge

Chuong Duong bridge is a major truss bridge that connects one of the most heavily trafficked routes in Vietnam-the 1A National road. Being built in the 80s of the twentieth century, after nearly 40 years of service, degradation and damages have threatened the integrity and safety of the structure. To evaluate the bridge structure, the first measurement campaign was carried out by the Faculty Civil Engineering of University of Transport and Communications (UTC) from the 10th–12th of May, 2021. During this campaign, measurement was conducted on the truss span number 10th with the total length of 89.28 m (Figs. 1 and 2).

During the measurement, eight high sensitivity accelerometers of type PCB were installed in each measurement set up. The full-scale operational modal analysis (OMA) test was performed by measuring the vibration under ambient excitation in 8

Fig. 1 Measurement span number 10th and its cross section

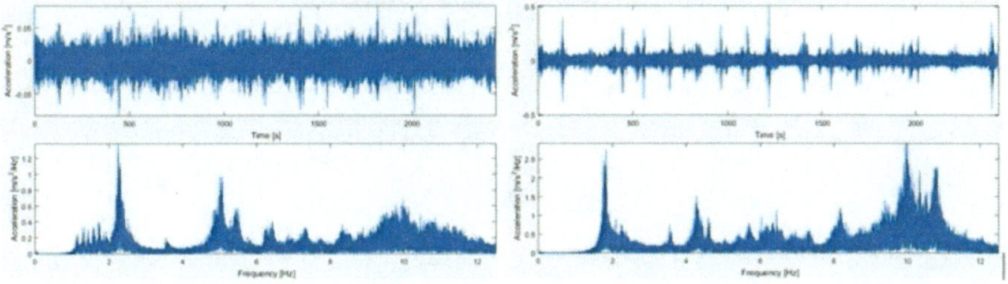

Fig. 2 Dynamic response of a transversal (left)—vertical (right) sensor in time/frequency domain

different setups during 30 min per each setup. The acquired data in time domain is first converted to frequency domain by Fast Fourier Transform (FFT), followed by being analyzed by covariance-driven Stochastic System identification (SSI-COV) method with MACEC program [14] to obtained the dynamic properties of the structure. The dynamic properties of the measured bridge is obtained and shown in Table 1.

Table 1 Dynamic properties obtained from measurement data

Mode number	Frequencies (Hz)	Mode type
1	1.79	Vertical bending
2	3.57	Torsion
3	4.30	Vertical bending
4	4.60	Lateral movement
5	5.03	Lateral bending

3.2 *Initial FE Model of Chuong Duong Bridge*

A FE model of the 10th truss span is built using the Stabil program [15]. This model (Fig. 3) consists of 68 nodes (including 1 reference node) and 191 beam elements with six degrees of freedoms (DOFs) at each node. The node includes translational displacements in X, Y, and Z—axes and the rotational displacements around X, Y and Z—axes. Pier P1 uses a pin bearing that only allows rotational displacements, whereas P2 applies a roller one that allows translational and rotational displacements. The truss structure is assumed to be simply supported and rigid connection in truss joints is assumed for the model. Non-structural components including maintenance paths, parapets, deck, etc., are evaluated as added masses at each node. The X-axis coincides with the longitudinal direction, while the Z-axis corresponds to the vertical direction and the Y-axis is in the horizontal axis of the bridge. Materials and truss member properties are detailed in Tables 2 and 3 respectively.

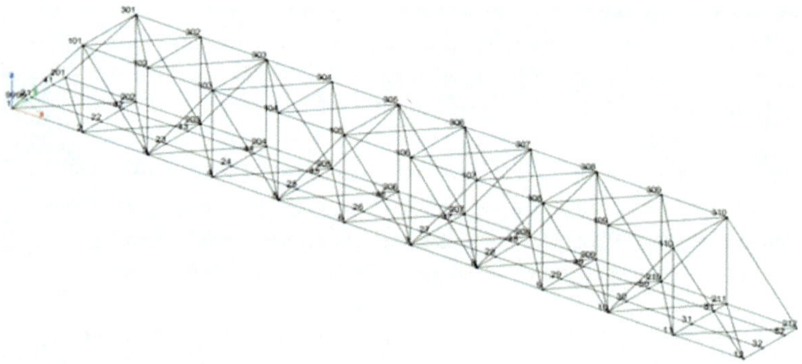

Fig. 3 FEM of Chuong Duong bridge

Table 2 Material properties of the FE model

Young's Modulus (GPa)	Poisson ratio	Density (Kg/m^3)
210	0.3	7850

Table 3 Properties of truss member

Truss member	Area (m^2)	Moment of Inertia Iz	Moment of Inertia Iy
Lateral bracing (bottom)	0.0080	0.00036	0.00065
Lateral bracing (top)	0.0081	0.00038	0.00071
Strut	0.067	0.01368	0.02194
Diagonal chord	0.067	0.01368	0.02194
Vertical chord	0.067	0.01368	0.02194

Table 4 Dynamic properties obtained from FE model

Mode number	Frequencies (Hz)	Mode type
1	1.85	Vertical bending
2	3.71	Torsion
3	4.61	Vertical bending
4	5.23	Lateral movement
5	5.47	Lateral bending

Dynamic properties of the FE model are obtained in Table 4 below.

3.3 *Model Updating Using BCMO*

For model updating of the bridge, 3 parameters were updated including the Young modulus, added mass m on the joints at the bottom chords of the bridge and added mass m_1 on the joints at the top chords of the bridge. Upper boundary (UB) and lower boundary (LB) of each parameter is represented in Table 5.

The objective function is then given by:

$$\text{Fitness} = \sum_{i=1}^{5} \frac{\omega_i 2}{\tilde{\omega}_i 2} = \sum_{i=1}^{5} (\omega_i - \tilde{\omega}_i)^2 / \tilde{\omega}_i^2 \tag{9}$$

where ω_i, $\tilde{\omega}_i$ are the natural frequencies of the simulated model and the experiment data respectively, i is the modal number. BCMO is applied to look for the convergence of the objective function. Convergence is shown in Fig. 4. For the optimization algorithm chosen, population size is 200 and the maximum number of iterations is 200. The procedure is repeated for GA, which is a widely used optimization algorithm to compare for the effectiveness of the proposed BCMO. The final results are shown in Table 6.

From the obtained results, we can see that BCMO provided a much better updating result for the modal frequencies of the five identified mode (especially in the 3rd mode) in compare with the initial FE model and the updated FE model using GA.

Table 5 LB and UB of updated parameters

	Young's modulus (GPa)	m(kg)	m_1(kg)
Lower boundary	190	25×10^5	5×10^3
Upper boundary	220	31×10^5	10×10^3

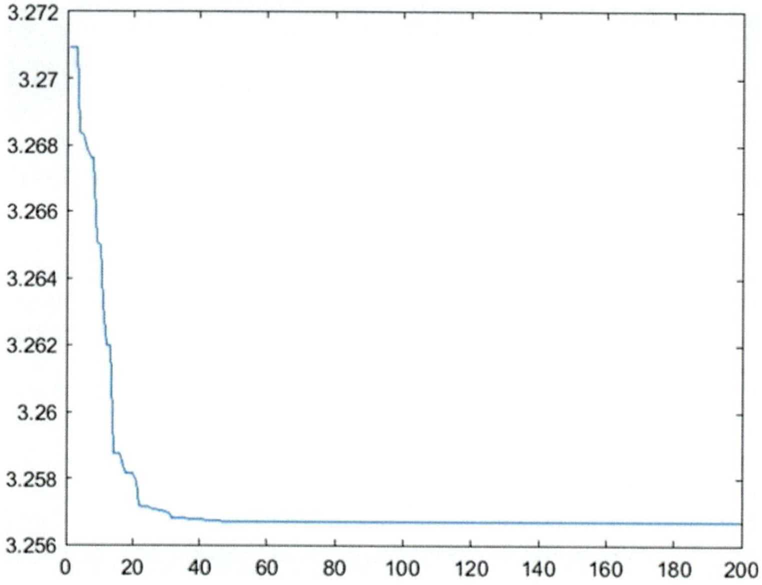

Fig. 4 Fitness tolerance of BCMO

Table 6 Comparison of modal frequencies from FE model after model updating and measurement

Mode number	Measurement (Hz)	Initial FE (Hz)	Updated FE-GA (Hz)	Updated FE-BCMO (Hz)
1	1.79	1.85 (3.35%)	1.73 (3.35%)	1.75 (2.23%)
2	3.57	3.71 (3.92%)	3.42 (4.20%)	3.45 (3.36%)
3	4.3	4.61 (7.2%)	4.25 (1.16%)	4.30 (0.00%)
4	4.6	5.23 (13.6%)	5.03 (9.34%)	4.90 (6.50%)
5	5.03	5.47 (8.74%)	5.36 (6.56%)	5.34 (6.10%)

4 Conclusion

In this paper, BCMO a recently developed metaheuristic optimization algorithm is applied for model updating of Chuong Duong truss bridge, which is a lifeline infrastructure in Vietnam. The result validated the superiority of the algorithm in term of accuracy when being compared with GA, which is another widely-used optimization algorithm. Since the vibration-based measurement data of Chuong Duong bridge evaluated in the paper is the first part of a long measurement campaign of the bridge, further investigation and improvement can be made from the next measurement data to achieve a better result for model updating as well as damage detection of the bridge.

Acknowledgements The authors acknowledge the financial support of VLIR-OUS TEAM Project, VN2018TEA479A103, 'Damage assessment tools for Structural Health Monitoring of Vietnamese infrastructures', funded by the Flemish Government.

vliruos SHARING MINDS, CHANGING LIVES With the support of THE BELGIAN DEVELOPMENT COOPERATION .be

References

1. Teughels A, De Roeck G (2004) Structural damage identification of the highway bridge Z24 by FE model updating. J Sound Vibrat 278(3):589–610
2. Jang S, Li J, Spencer B (2013) Corrosion estimation of a historic truss bridge using model updating. J Bridg Eng 18. https://doi.org/10.1061/(ASCE)BE.1943-5592.0000403
3. Sierra M, Coello C (2005) Improving PSO-based multi-objective optimization using crowding, mutation and ∈-dominance. Lect Notes Comput Sci 3410. https://doi.org/10.1007/978-3-540-31880-4_35
4. Altiparmak F, Gen M, Lin L, Paksoy T (2006) A genetic algorithm approach for multi-objective optimization of supply chain networks. Comput Ind Eng 51:196–215. https://doi.org/10.1016/j.cie.2006.07.011
5. Mirjalili S, Mirjalili SM, Lewis A (2014) Grey wolf optimizer. Adv Eng Softw 69:46–61, ISSN 0965-9978. https://doi.org/10.1016/j.advengsoft.2013.12.007
6. Mirjalili S, Lewis A (2016) The whale optimization algorithm. Adv Eng Softw 95:51–67, ISSN 0965-9978, https://doi.org/10.1016/j.advengsoft.2016.01.008
7. Mirjalili S (2015) Moth-flame optimization algorithm: a novel nature-inspired heuristic paradigm. Knowl-Based Syst 89:228–249, ISSN 0950-7051. https://doi.org/10.1016/j.knosys.2015.07.006
8. Faramarzi A, Heidarinejad M, Mirjalili S, Gandomi A (2020) Marine predators algorithm: a nature-inspired metaheuristic. Expert Syst Appl 152:113377. https://doi.org/10.1016/j.eswa.2020.113377
9. Tiachacht S, Khatir S, Le Thanh C, Venkata Rao R, Mirjalili S, Abdel Wahab M, Inverse problem for dynamic structural health monitoring based on slime mould algorithm. Engineering with Computers. https://doi.org/10.1007/s00366-021-01378-8. (2021).
10. Tran-Ngoc H, Khatir S, De Roeck G, Bui-Tien T, Nguyen-Ngoc L (2018) Abdel WM Model updating for nam o bridge using particle swarm optimization algorithm and genetic algorithm. Sensors 18(12):4131
11. Ho, VL, De Roeck G, Bui Tien T, Wahab M (2021) Determination of the effective stiffness of half-open cross-section bars and orthotropic steel deck of a truss bridge using model updating. https://doi.org/10.1007/978-981-15-9893-7_6
12. Nguyen-Ngoc L, Tran-Ngoc H, Nguyen-Tran H, Nguyen-Duc B, Nguyen-Le-Minh D, Bui-Tien T, Wahab MA (2021) Damaged detection in structures using artificial neural networks and genetic algorithms. In: *Proceedings of the 3rd international conference on sustainability in civil engineering. Lecture notes in civil engineering*, vol 145. Springer, Singapore. https://doi.org/10.1007/978-981-16-0053-1_4
13. Le-Duc T, Nguyen, Q-H, Nguyen-Xuan H (2020) Balancing composite motion optimization. Inf Sci 520. https://doi.org/10.1016/j.ins.2020.02.013

14. Reynders E, Schevenels M, Roeck GD (2011) MACEC 3.2: a matlab toolbox for experimental and operational modal analysis; user's manual; Report BWM-2011-XX; Leuven University, Leuven, Belgium
15. Dooms D, Jansen M, De Roeck G, Degrande G, Lombaert G, Schevenels M, François S (2010) StaBIL: a finite element toolbox for Matlab, 2.0 ed. Structural Mechanics Section of the Department of Civil Engineering, KU Leuven, Leuven, Belgium

A Two-Step Approach for Damage Detection in a Real 3D Tower Using the Reduced-Order Finite Element Model Updating and Atom Search Algorithm (ASO)

Hoang-Le Minh, Thanh Sang-To, Tran-Thanh Danh, Nguyen-Ngoc Phu, Magd Abdel Wahab, and Thanh Cuong-Le

Abstract In this paper, an effective approach is presented for solving damage identification problems. The key factor of this approach is based on the model updating and inverse method. First, Root-Mean-Square-Error (RMSE), which registers the differences between frequencies at two states; damaged stage and undamaged stage, is used to establish the objective function. Then, an effective optimization algorithm named the Atom search algorithm (ASO) is employed to minimize the objective function, which accounts for variables related to stiffness reduction in the structures. The process of calculating the objective function during the course of iterations is based on the reduced-order finite element model of a real 3D tower named Guangzhou New TV Tower (GNTT) to extract the frequencies at each iteration. This process will implement continuously until finding the damaged values that agree with the objective function with the minimum error. The results obtained in this paper prove that the proposed approach can predict damaged stories with high reliability and acceptance error.

Keywords Structural damage detection · Natural frequency · Finite element method · Atom search algorithm · Model updating

1 Introduction

Structural Health Monitoring (SHM) is a significant factor in research over the past two decades because assessing the location and extent of damaged elements in the structure will ensure effective working of the structure and timely repair. The inverse method combined with model updating is a reliable method to predict damaged

H.-L. Minh (✉) · M. Abdel Wahab
Department of Electrical Energy, Metals, Mechanical Constructions, and Systems, Faculty of Engineering and Architecture, Ghent University, 9000 Gent, Belgium
e-mail: MinhHoang.Le@Ugent.be

H.-L. Minh · T. Sang-To · T.-T. Danh · N.-N. Phu · T. Cuong-Le
Faculty of Civil Engineering, Ho Chi Minh City Open University, Ho Chi Minh City, Vietnam

M. Abdel Wahab (ed.), *Proceedings of the 2nd International Conference on Structural Damage Modelling and Assessment*, Lecture Notes in Civil Engineering 204,
https://doi.org/10.1007/978-981-16-7216-3_2

elements in SHM. This approach has been used widely and successfully in recent studies [1–3]. Friswell [4] introduced a brief overview of applying the model updating technique and inverse methods for SHM from vibration data obtained from measurements. The main features of this method include three steps: (i) establish the objective function, (ii) the model updating technique is used to calculate the objective function, and (iii) finding a reliable algorithm to minimize the objective function. Thus, the selection of a suitable optimization algorithm takes a crucial role in the successful method. The robust development of metaheuristic algorithms with different inspirations has brought many choices to solve optimization problems. There are many algorithms such as Atom search algorithm (ASO) [5], Cuckoo search (CS) [6], Gravitational Search Algorithm (GSA) [7], Black Hole [8], Grey Wolf optimizer (GWO) [9], The Whale Optimization Algorithm (WOA) [10], and so on. These algorithms have proven effective in their application to SHM. Minh et al. [11] presented a new model updating technique combined with an improved PSO named (EHVPSO) for damage detection in a real 3D structure. In this study, a real 3D transmission tower was used to prove the effectiveness of the method. The results obtained in this study prove the reliability and high level. Cuong-Le et al. [12] utilize PSO to enhance the ability of support vector machine in damage detection. Alkayem [13] used a finite element (FE) model built in MATLAB and some algorithms, including Particle Swarm Optimization (PSO), and Differential Evolution (DE) and Genetic Algorithm (GA), to detect damaged elements in 3D frame structures. Chen [14] proposed a new method for calculating the objective function, then the new objective function collaborated with Whale optimization algorithm (WOA) to predict the damaged elements in a simply-supported beam and 31-bar truss structures. Nozari et al. [15] used a FE model updating for damage detection of 10-story building using ambient vibration measurements, etc. Almost all previous public studies in this field concerned with simple structures including 2D or 3D frames.

In this paper, to assess generally this method, a real structure named Guangzhou New TV Tower located China is used to detect the damaged elements. First, For simplicity and to reduce the number of degrees of freedom of the structure, a reduced-order FE model generated from the full-order model is conducted by MATLAB. Then, a recent optimization algorithm named Atom search optimization (ASO) is employed as a reliable algorithm to detect damaged elements in this structure. The damaged detection process will be secured by the exchange data between the FE model and the ASO algorithm. This process will stop if the objective function achieves a suitable convergence rate and at the same time, these variants in the objective function will determine the extent of damage in the structure.

2 Structural Modeling of the Guangzhou New TV Tower

In this section, a real 3D tower named the Guangzhou New TV Tower is selected to validate the proposed method. The system of this structure is tube-in-tube with 600 height. The tower includes two main parts; the first part is the main tower with 454 m

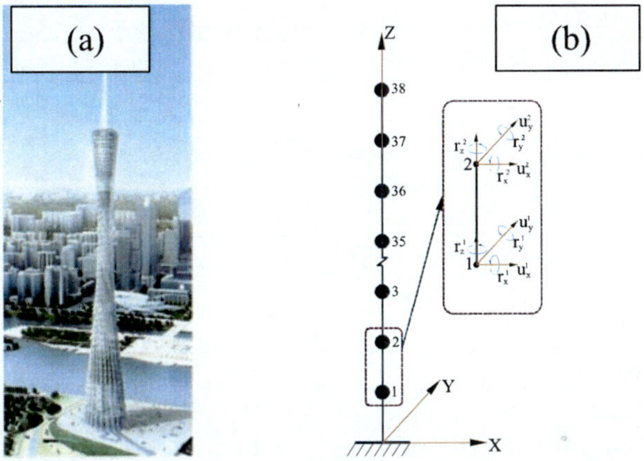

Fig. 1 **a** The Guangzhou New TV Tower according to Chen et al., **b** a reduced-order FE model

height, and the second part is the antenna mast with 146 m height, according to Chen et al. [16] as shown in Fig. 1a. Because of the complicated geometry of the structure, using the full-scale FE model will have difficulties in this study. To simplify the model, a reduced-order FE model is simulated in a simple way. Thus, the simple FE model includes 37 elements; each element is simulated as a linear elastic beam element. Consequently, the FE model is modelled as a cantilever beam including 37 elements. Each element registers two nodes at start and end, respectively. The total number of nodes is 38, the label of nodes will in a gradual increase from 1 to 38, the vertical coordinates of each node are given in Table 1. In the reduced model, the mass of each story is lumped at a node that is located at the same high level of the story mentioned, then these masses are lumped to the adjacent reducing center, as shown in Fig. 1b. The vertical displacement following the Z direction is ignored in the FE reduced model. Thus, each node registers 5 DOFs, including two horizontal translational DOFs and three rotational DOFs, as shown in Fig. 1b. Thus, each element has 10 DOFs and will register stiffness matrix with dimensions 10 × 10. The FE model in this study is referred to the model presented by Ni, Xia [21] and Chen, Lu [20]. Thus, each element stiffness matrix is given in Eq. (1).

$$K_i^e = \begin{bmatrix} K^e_{11} & K^e_{12} & K^e_{13} & K^e_{14} & K^e_{15} & K^e_{16} & K^e_{17} & K^e_{18} & K^e_{19} & K^e_{110} \\ & K^e_{22} & K^e_{23} & K^e_{24} & K^e_{25} & K^e_{26} & K^e_{27} & K^e_{28} & K^e_{29} & K^e_{210} \\ & & K^e_{33} & K^e_{34} & K^e_{35} & K^e_{36} & K^e_{37} & K^e_{38} & K^e_{39} & K^e_{310} \\ & & & K^e_{44} & K^e_{54} & K^e_{46} & K^e_{47} & K^e_{48} & K^e_{49} & K^e_{410} \\ & & & & K^e_{55} & K^e_{56} & K^e_{57} & K^e_{58} & K^e_{59} & K^e_{510} \\ & & & & & K^e_{66} & K^e_{67} & K^e_{68} & K^e_{69} & K^e_{610} \\ & & & & & & K^e_{77} & K^e_{78} & K^e_{79} & K^e_{710} \\ & & & Sym & & & & K^e_{88} & K^e_{89} & K^e_{810} \\ & & & & & & & & K^e_{99} & K^e_{910} \\ & & & & & & & & & K^e_{1010} \end{bmatrix} \quad (1)$$

Table 1 The nodal coordinate (z) of the reduced model

Floor	Label of node	Z (m)	Floor	Label of node	Z (m)
F1	1	−10	F21	21	375.85
F2	2	0	F22	22	381.2
F3	3	12	F23	23	396.65
F4	4	22.25	F24	24	407.05
F5	5	27.6	F25	25	417.45
F6	6	58.65	F26	26	427.85
F7	7	84.65	F27	27	438.25
F8	8	95.05	F28	28	443.6
F9	9	105.45	F29	29	480
F10	10	116.2	F30	30	497
F11	11	147.05	F31	31	505.2
F12	12	157.45	F32	32	520.7
F13	13	168	F33	33	531.2
F14	14	204.25	F34	34	545.2
F15	15	225.2	F35	35	565.2
F16	16	272	F36	36	580.7
F17	17	308.25	F37	37	598
F18	18	329.2	F38	38	610
F19	19	344.64			
F20	20	355.05			

3 The Objective Functions

The objective function is selected from the correlation of natural frequencies obtained from the FE model with cases of damage and data measured from testing of healthy structure. By solving Eq. (2) using the FE model, we can get the natural frequencies.

$$\left[K - \omega_i^2 M\right]\{\phi_i\} = \{0\} \tag{2}$$

where $\boldsymbol{K}$ and $\boldsymbol{M}$ are the structural stiffness matrix and mass matrix, respectively. ω_i is natural period and ϕ_i is mode shape vector.

From ω_i, we can calculate the frequencies as follows Eq. (3):

$$T_i = \frac{2\pi}{\omega_i},\ f_i^{FE} = \frac{1}{T_i} \tag{3}$$

The objective function is selected by using Root-Mean-Square-Error (RMSE) as shown in Eq. (4).

$$f_{ofun} = \frac{\sqrt{\sum_{i=1}^{n} \left(f_i^{Measured} - f_i^{FEM}\right)^2}}{n} \tag{4}$$

where n is the number of frequencies obtained using the FE model.

4 Atom Search Algorithm (ASO)

ASO [5] is inspired by basic molecular dynamics. From this perspective, each position X_i $(i = 1, 2, ..., N)$ is considered as a candidate solution in an unknown search space dimension.

4.1 *Interaction Force in ASO*

The interaction force obtained from the Lennard–Jones (L-J) potential is a simple mathematical revised to obtain more positive attraction and less negative repulsion as iterations increase. The interaction force impact on the ith Atom from jth Atom at tth iteration is shown in Eq. (5)

$$\begin{aligned} F_{ij}^{d} &= -\eta(t)\left[2\left(h_{ij}(t)\right)^{13} - \left(h_{ij}(t)\right)^{7}\right] \\ \eta(t) &= \alpha\left(1 - \frac{t-1}{T_{\max}}\right)^{3} e^{-\frac{20t}{T_{\max}}}, \quad \alpha = [10,\ 20,\ 30, ...,\ 100] \end{aligned} \tag{5}$$

where α is the depth weight, t is the current of iteration, and $T_{\max}$ is the maximum of iterations.

$h_{ij}(t)$ is the ratio of the distance between two ith and jth atoms to the length scale $\sigma(t)$ as given in Eq. (6).

$$h_{ij}(t) = \begin{cases} h_{\min} & \frac{\left\|r_{ij}(t)\right\|_2}{\sigma(t)} < h_{\min} \\ \frac{\left|r_{ij}(t)\right|}{\sigma(t)} & h_{\min} \le \frac{\left\|r_{ij}(t)\right\|_2}{\sigma(t)} \le h_{\max} \\ h_{\max} & \frac{\left\|r_{ij}(t)\right\|_2}{\sigma(t)} > h_{\max} \end{cases} \tag{6}$$

where $h_{\max}$ and $h_{\min}$ are lower and upper boundaries, respectively and defined as Eq. (7).

$$\begin{cases} h_{\min} = g_0 + g(t) \\ h_{\max} = u \end{cases} = \begin{cases} h_{\min} = g_0 + 0.1 \sin\left(\frac{\pi t}{2T_{\max}}\right) \\ h_{\max} = u \end{cases} \tag{7}$$

$\sigma(t)$ is denoted the length scale, and it can be expressed by Eq. (8)

$$\begin{aligned} \sigma(t) &= \left\| X_{ij}(t), X_{Kbest}(t) \right\|_2 \\ X_{Kbest}(t) &= \frac{\sum_{j \in K_{best}} X_{ij}(t)}{K(t)} \end{aligned} \tag{8}$$

where K_{best} can be defined as a high-reliability search space for next iterations, and it is described by Eq. (9).

$$K_{best}(t) = N - (N - 2)\sqrt{\frac{t}{T_{\max}}} \tag{9}$$

4.2 Mathematical Representation of Geometric Constraint and the Mass of Atom

To increase the convergence rate, each Atom is linked with the best Atom. In other words, each Atom is affected by the best Atom through the force called geometric constraint force shown in Eq. (10).

$$\begin{aligned} & G_i^d(t) = \lambda(t)\left[X_{best}^d(t) - X_i^d(t)\right] \\ & \lambda(t) = \beta e^{-\frac{20t}{T_{\max}}} \\ & \beta = [0.1;\ 0.2;\ 0.3;\ 0.4;\ 0.5;\ 0.6;\ 0.7;\ 0.8;\ 0.9;\ 1] \end{aligned} \tag{10}$$

where $X_{best}^d(t)$ is the best solution at tth iteration, $\lambda(t)$ is the Lagrangian multiplier.

In ASO, each Atom $X_i^d(t)$ registers a changeable mass and can be expressed in Eq. (11).

$$\begin{aligned} M_i(t) &= e^{-\frac{(Fit_i(t) - Fit_{best})}{Fit_{worst} - Fit_{best}}} \\ m_i(t) &= \frac{M_i(t)}{\sum_{j=1}^{N} M_j(t)} \end{aligned} \tag{11}$$

where $Fit_{best}(t)$ and $Fit_{worst}(t)$ are the maximum and minimum values of the objective function at the tth iteration, respectively. N is the number of Atom, $Fit_i(t)$ is the value of the objective function of Atom ith at tth iteration. $Fit_{best}(t)$ and $Fit_{worst}(t)$ are expressed as Eq. (12).

$$\begin{aligned} Fit_{best}(t) &= \min_{(i=1,2,\dots,N)} [Fit_i(t)] \\ Fit_{worst}(t) &= \max_{(i=1,2,\dots,N)} [Fit_i(t)] \end{aligned} \tag{12}$$

4.3 Atomic Motion

The process of acceleration updating is calculated by Eq. (13).

$$\begin{aligned} a_i^d(t) &= \frac{F_i^d(t)}{m_i^d(t)} + \frac{G_i^d(t)}{m_i^d(t)} \\ &= \alpha\left(1 - \frac{t-1}{T_{\max}}\right)^3 e^{-\frac{20t}{T_{\max}}} \sum_{j \in K_{best}} \frac{rand_j\left[2\left(h_{ij}(t)\right)^{13} - \left(h_{ij}(t)\right)^7\right]}{m_i(t)} \frac{\vec{r_{ij}}}{\left\|r_{ij}\right\|_2} \\ &+\beta e^{-\frac{20t}{T_{\max}}} \frac{\left(X_{best}^d(t) - X_i^d(t)\right)}{m_i(t)} \end{aligned} \tag{13}$$

where $\vec{r_{ij}}(t)$ and $\left\|r_{ij}(t)\right\|_2$ are the position difference vector and Euclidean distance between the ith and the jth Atoms, respectively and can be expressed by Eq. (14).

$$\vec{r_{ij}}(t) = \left(X_j^d(t) - X_i^d(t)\right), \quad \left\|r_{ij}(t)\right\|_2 = \sqrt{\sum_{k=1}^{D} (x_{jk} - x_{ik})^2} \tag{14}$$

The position updating of each Atom is updated through the velocity updating process and can be written as Eqs. (15 and 16).

$$V_i^d(t+1) = rand_i^d V_i^d(t) + a_i^d(t) \tag{15}$$

$$X_i^d(t+1) = X_i^d(t) + V_i^d(t+1) \tag{16}$$

5 Application to Structural Health Monitoring

5.1 Structural Damage Identification Approach

The damaged identification can be illustrated by a scalar vector $X_i = \left(x_{1,}x_2, ..., x_n\right)$ with each individual x_i $(i = 1, 2, ..n)$ bounded in the ranged [0, 1]. Thus, the global stiffness matrix will reduce at damaged stage and given in Eq. (17).

$$[K] = \sum_{i=1}^{n} (1 - x_i)k_i; \quad 0 \leq x_i \leq 1 \tag{17}$$

where k_i is the stiffness matrix of element ith at the healthy stage.

The goal of structural damage identification is to determine the scalar vector $X_i = \left(x_{1,}x_2, ..., x_n\right)$, which agrees with the objective function with acceptable error. The process of detecting $X_i = \left(x_{1,}x_2, ..., x_n\right)$ is secured by an optimization algorithm. In this paper, ASO is employed to do this. And the process of damage identification using FE model updating and ASO is illustrated in Fig. 2.

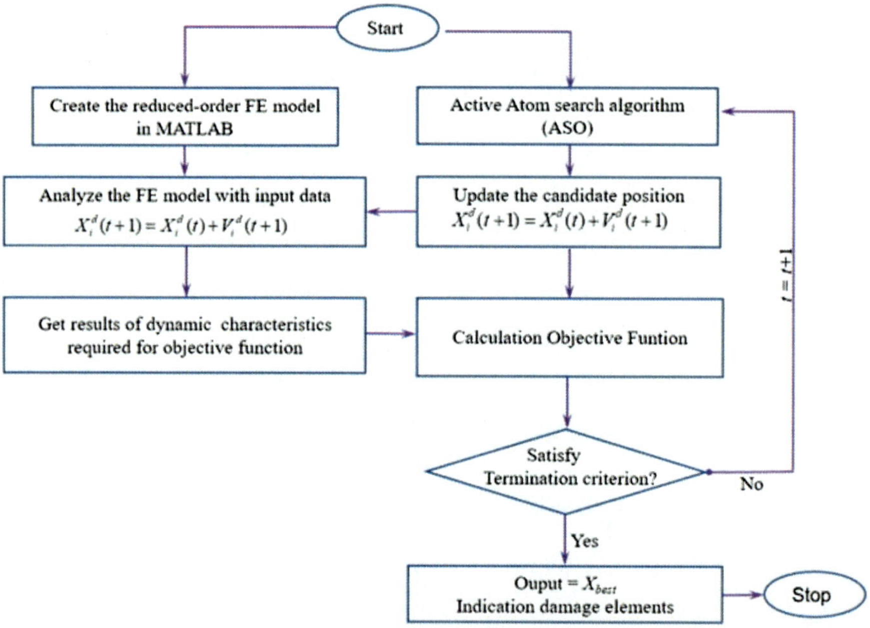

Fig. 2 The process of detecting damaged structures using a reduced-order FE model and Atom search algorithm

Table 2 Reduction in stiffness in stories of Guangzhou New TV Tower for different damage cases

Case study	Damaged story	Damaged severity (%)
Case 1	Story 01 Story 10	$x_{65} = 35$ $x_{10} = 50$
Case 2	Story 02 Story 12 Story 17	$x_{02} = 20$ $x_{05} = 10$ $x_{10} = 25$

5.2 *Application to Guangzhou New TV Tower*

To demonstrate the reliability of the proposed method, two damaged cases with different reductions of stiffness of each story are considered as given in Table 2. The value of frequencies at two stages damaged and undamaged are shown in Table 3.

Table 3 The values of the 40 first frequencies using the reduced-FE model at undamaged stage and two cases of damaged stage

Mode	Frequency			Mode	Frequency		
	undamaged structure	damaged Case 1	damaged Case 2		undamaged structure	damaged Case 1	damaged Case 2
1	0.1104	0.107	0.1234	21	2.6741	2.6386	2.2999
2	0.1587	0.1633	0.1319	22	2.8391	2.7123	2.4595
3	0.3463	0.3391	0.1885	23	2.9926	2.8829	2.5209
4	0.3688	0.3704	0.4027	24	3.1085	3.0383	2.5653
5	0.3994	0.3865	0.4138	25	3.1542	3.1076	2.6515
6	0.4605	0.4541	0.4273	26	3.2686	3.1817	2.8079
7	0.485	0.4763	0.4373	27	3.3336	3.2465	2.8952
8	0.7381	0.5977	0.5187	28	3.4356	3.3398	2.9331
9	0.9026	0.7702	0.6611	29	3.4725	3.4125	2.968
10	0.9972	0.9848	0.9138	30	3.8553	3.5174	3.1053
11	1.0373	1.016	0.9433	31	3.967	3.5822	3.2253
12	1.1218	1.0838	0.9959	32	4.1373	3.9573	3.5532
13	1.2436	1.1958	1.0593	33	4.191	4.1902	3.6583
14	1.5031	1.4395	1.2325	34	4.2603	4.2592	3.7393
15	1.7261	1.6523	1.431	35	4.3642	4.2804	3.8077
16	1.8051	1.7219	1.4944	36	4.713	4.363	4.2691
17	1.9809	1.9669	1.5112	37	4.7321	4.713	4.4802
18	1.995	1.9897	1.5436	38	5.2113	4.9201	4.5175
19	2.1916	2.0908	1.8409	39	5.3721	5.2079	4.6543
20	2.3632	2.2616	1.9492	40	5.6646	5.3282	4.7345

The results of the damaged indicator process will be presented according to the following:

- *The convergence trend of objective function*: These curves show the trend of the value of the objective function over the course of iterations.
- *The historical trend of damaged elements*: These curves show the changes in the values of the damaged variables, which are defined as a reduction of stiffness. They also illustrate the fast or slow convergence rate of ASO algorithm.
- *The damage identification bar chart*: The final values are shown in the bar chart. The chart shows the correlation between the values of two damaged cases and damaged indicators using FE model updating combined with ASO.

The results are shown from Figs. 3, 4, 5, 6, 7, and 8 and the statistical table in true value and indicator value is given in Table 4.

Fig. 3 The convergence trend of objective function in Case 1

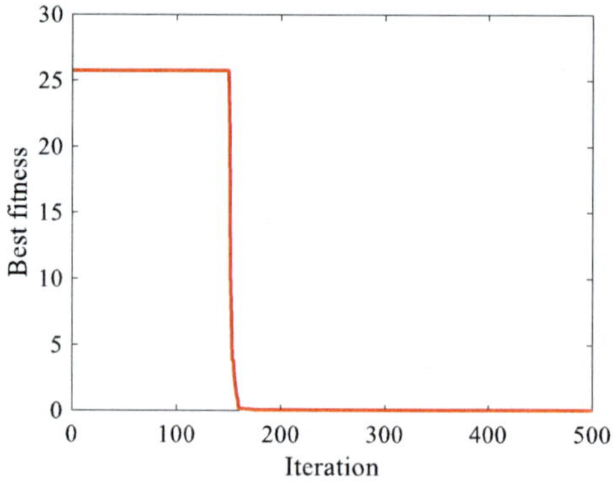

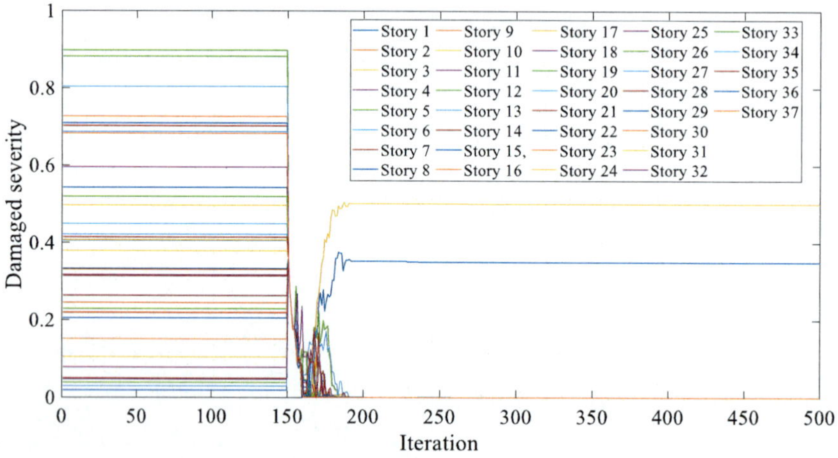

Fig. 4 The historical trend of damaged elements in Case 1

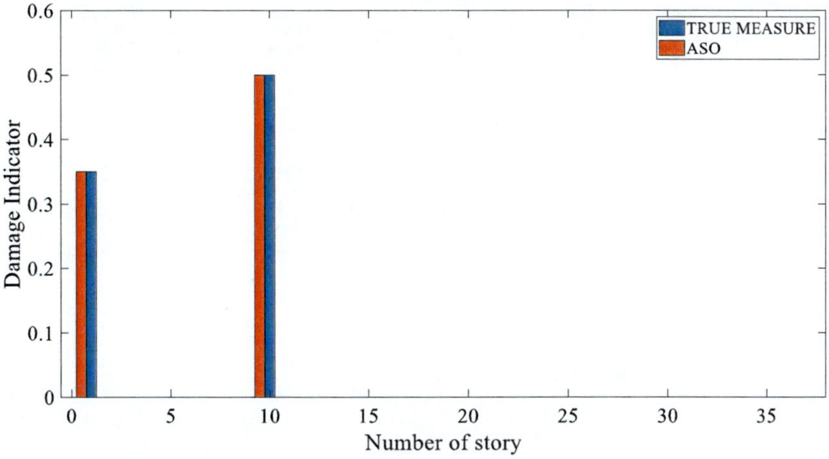

Fig. 5 The damage identification bar chart in Case 1

Fig. 6 The convergence trend of objective function in Case 2

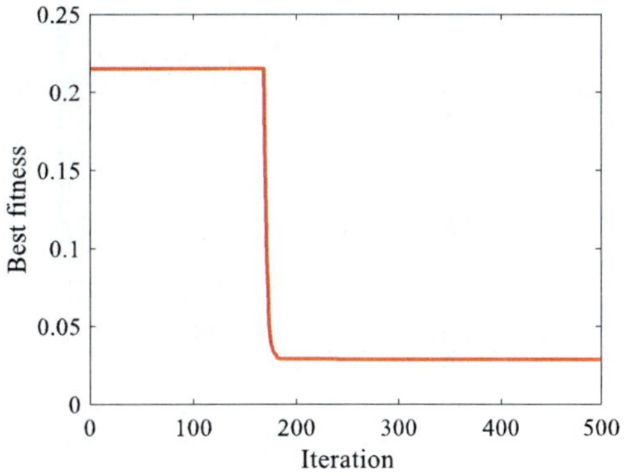

The results in Table 4 show that ASO can predict accurately the damaged stories in Case 1 and acceptable errors in Case 2. There is a general acceptance that the ability to explore ASO during the first few iterations does not appreciate. However, this ability is improved more clearly during the last iterations, and ASO achieves stability in convergence rate and accuracy level because of reducing the Lennard-Jones (L-J) potential force during the last iterations.

6 Conclusion

The paper introduced an effective method to predict damaged structures for a real complex structure named Guangzhou New TV Tower. In the paper, a complex model structure can be simplified by a simple model structure in which each floor is lumped

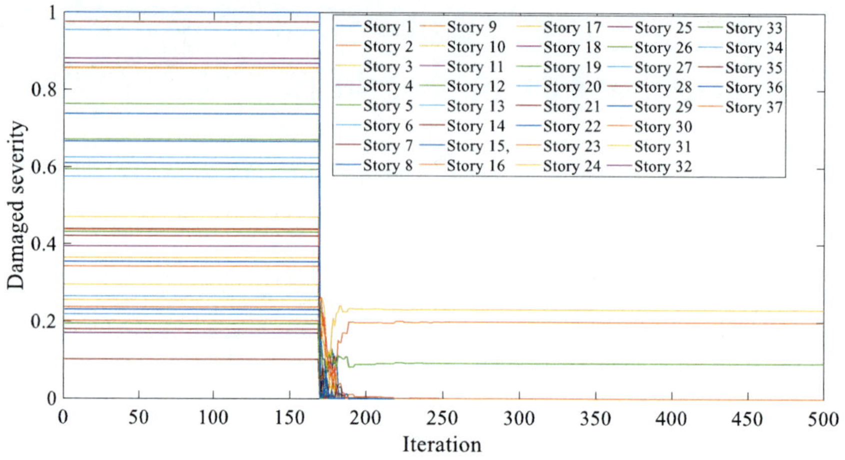

Fig. 7 The historical trend of damaged elements in Case 2

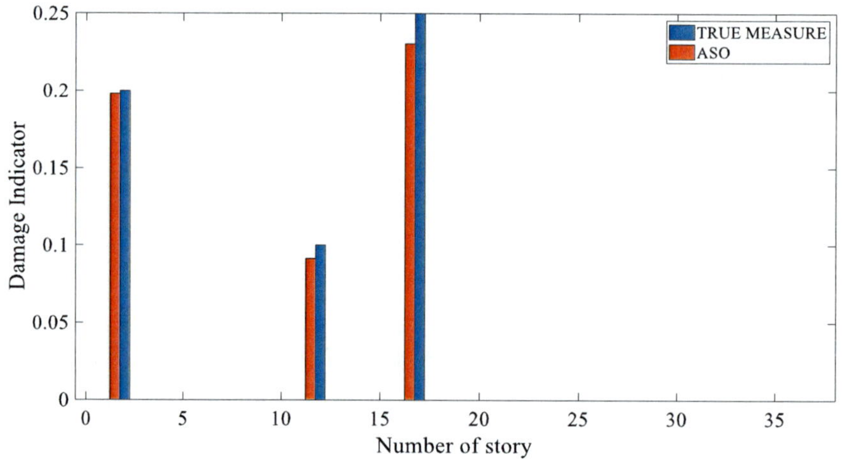

Fig. 8 The damage identification bar chart in Case 2

Table 4 The results of ASO algorithms for predicting damage elements

Case study	Story	True measure	ASO algorithm
			Damaged indicator
Case 1	Story 01	0.35	0.35
	Story 10	0.5	0.5
Case 2	Story 02	0.2	0.198
	Story 12	0.1	0.0914
	Story 17	0.25	0.2304

at the same level as the real floor. The stiffness of each story is converted to the stiffness of the frame whose registers ten degrees of freedom. The process of damaged identification is performed using the FE model updating and inverse method. The results obtained in this paper prove the effectiveness of the method and the reliability of the ASO algorithm. However, this method still has limitations because it cannot predict the level of damage severity of the individual elements on each floor. This method can be used as a reference method to assess the level of damage story quickly. For a more detailed assessment, we need to apply a new model updating technique with a full-scale model, which can be simulated by finite element software.

Acknowledgements The authors acknowledge the financial support of VLIR-UOS TEAM Project, VN2018TEA479A103, 'Damage assessment tools for Structural Health Monitoring of Vietnamese infrastructures' funded by the Flemish Government.

The authors gratefully acknowledge the financial support granted by the Scientific Research Fund of the Ministry of Education and Training (MOET), Vietnam (No. B2021-MBS-06).

References

1. Alkayem NF et al (2018) Structural damage detection using finite element model updating with evolutionary algorithms: a survey. Neural Comput Appl 30(2):389–411
2. Alkayem NF, Cao M, Ragulskis M (2018) Damage diagnosis in 3D structures using a novel hybrid multiobjective optimization and FE model updating framework. Complexity
3. Ding Z, Li J, Hao H (2019) Structural damage identification using improved Jaya algorithm based on sparse regularization and Bayesian inference. Mech Syst Signal Process 132:211–231
4. Friswell M, Penny J (1992) A simple nonlinear model of a cracked beam. In proceedings of the international modal analysis conference. SEM Society for Experimental Mechanics Inc.
5. Zhao W, Wang L, Zhang Z (2019) Atom search optimization and its application to solve a hydrogeologic parameter estimation problem. Knowl-Based Syst 163:283–304
6. Yang X-S, Deb S (2009) Cuckoo search via Lévy flights. In: 2009 World congress on nature & biologically inspired computing (NaBIC). IEEE
7. Rashedi E, Nezamabadi-pour H, Saryazdi S (2009) GSA: a gravitational search algorithm. Inf Sci 179(13):2232–2248
8. Hatamlou A (2013) Black hole: a new heuristic optimization approach for data clustering. Inf Sci 222:175–184
9. Mirjalili S, Mirjalili SM, Lewis A (2014) Grey wolf optimizer. Adv Eng Softw 69:46–61
10. Mirjalili S, Lewis A (2016) The whale optimization algorithm. Adv Eng Softw 95:51–67
11. Minh H-L et al (2021) An Enhancing Particle Swarm Optimization Algorithm (EHVPSO) for damage identification in 3D transmission tower. Eng Struct 242:112412
12. Cuong-Le T et al (2021) An efficient approach for damage identification based on improved machine learning using PSO-SVM. Eng Comput
13. Alkayem NF, Cao M (2018) Damage identification in three-dimensional structures using single-objective evolutionary algorithms and finite element model updating: evaluation and comparison. Eng Optim 50(10):1695–1714

14. Chen Z, Yu L (2020) A novel WOA-based structural damage identification using weighted modal data and flexibility assurance criterion. Struct Eng Mech 75(4):445–454
15. Nozari A et al (2017) Effects of variability in ambient vibration data on model updating and damage identification of a 10-story building. Eng Struct 151:540–553
16. Chen W et al (2011) Theoretical and experimental modal analysis of the Guangzhou New TV Tower. Eng Struct 33(12):3628–3646

Monitoring Bridge Frequencies Using Passing Vehicle

Duong Huong Nguyen, Quoc Bao Nguyen, and Magd Abdel Wahab

Abstract In this paper, acceleration responses from a vehicle passing through the bridge are used to monitor the bridge frequencies. A bridge and vehicle system is set up in the laboratory. Sensors are attached to both bridge (direct method) and vehicle (indirect method) to compare the results. The damage is introduced in the bridge to observe the differences in the bridge frequencies between direct and indirect method. The modal analysis method called Covariance-based Stochastic System Identification (SSI) is chosen to analyse the response data. The natural frequencies of both intact and damaged bridge are extracted. Comparing the experimental results between direct and indirect methods shows that using a passing vehicle could identify the bridge frequency.

Keywords Structural Health Monitoring (SHM) · Bridge frequencies estimation · Vehicle–bridge interaction · Indirect method

1 Introduction

The modal properties of a bridge such as natural frequencies, mode shapes are the key parameters for Structural Health Monitoring (SHM). Many works have been done to analyse those parameters and use for SHM. Sensors attached on the bridge to collect the vibration response data and monitor the dynamic properties are usually used. Those methods rely directly on the bridge response, therefore can be referred to as the direct method. The vehicle is only considered as an excitation force [1–4]. A number of vibration sensors needs to be attached to the single bridge, especially for mode shapes' identification. Moreover, the experimental campaign is hard to perform, costly, and impossible to transfer to other bridges.

D. H. Nguyen (✉) · Q. B. Nguyen
Faculty of Bridges and Roads, National University of Civil Engineering, 100000 Hanoi, Vietnam
e-mail: duongnh2@nuce.edu.vn

M. Abdel Wahab
Soete Laboratory, Department of Electrical Energy, Metals, Mechanical Constructions and Systems, Faculty of Engineering and Architecture, Ghent University, 9000 Gent, Belgium

M. Abdel Wahab (ed.), *Proceedings of the 2nd International Conference on Structural Damage Modelling and Assessment*, Lecture Notes in Civil Engineering 204,
https://doi.org/10.1007/978-981-16-7216-3_3

Recently, researchers have been focused on the indirect method for bridge measurement. Sensors are attached to the vehicle to monitor the modal parameters of the bridge. Therefore, only a small number of sensors is needed. The idea of using a passing vehicle to extract the bridge frequency was attempted by Yang et al. in 2004 [5] and verified by a field test one year later [6]. The mode shapes of bridge also can be found with limited sensors [7]. Research has been done to prove that the frequencies of a bridge can be identified using indirect method [5, 8–10]. The vibration response of passing vehicle also can be used for damage detection purposes. Yang et al. [11] proposed a new damage index calculated from the acceleration of a passing vehicle. This research proved that the indirect method was more accurate in the noisy environment than the direct method. The state of art review on damage detection using a passing vehicle can be found in Ref. [12].

In this paper, a laboratory bridge was set up to verify the application of the indirect method for damage detection. The experiment proves that the vibration response of the passing vehicle contains the information of the bridge vibration. Therefore, using the indirect method in SHM seems to be a promising method that has low cost, mobility, and efficiency.

2 Modal Frequency Extraction from the Theoretical Response of a Vehicle Passing Over a Bridge

Consider a vehicle passing over a simply supported bridge in Fig. 1. The vehicle with speed ϑ is modeled as a sprung mass m_v and stiffness k_v. q_v is the vertical displacement of the strung mass and q_b is the deflection of the bridge. ρ, A, E, I, L are the density, sectional area, Young's modulus, moment of inertia, and span of the beam, respectively. The damping of the bridge and vehicle are both neglected. The bridge is modelled as a simply supported beam with a smooth road surface. The equation of motion for the passing vehicle is given by [5]:

$$m_v\ddot{q}_v + k_v(q_v - q_b(x = \vartheta t)) = 0 \tag{1}$$

where $q_b(x = \vartheta t)$ is the deflection of the beam at the strung mass location.

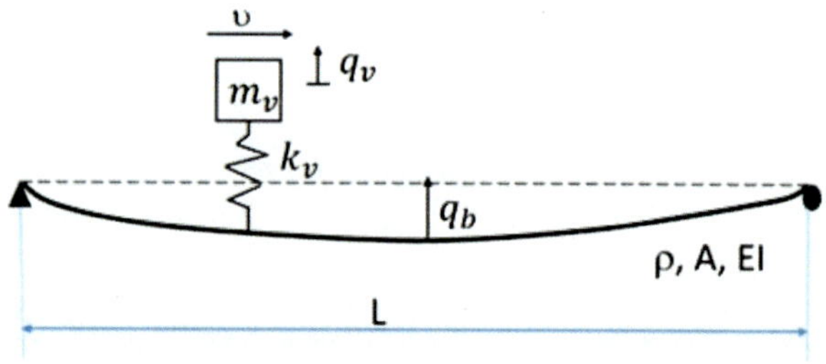

Fig. 1 Passing vehicle over a beam

The vehicle response can be obtained by assuming the vehicle mass is much less than the mass of bridge and considering only the first mode of the beam. Moreover, the contact force between strung mass is considered and the beam and the displacement $u_{x=\vartheta t}$ of the beam is approximated to derive the equation. The established procedure is presented by Yang et al. [5].

$$\ddot{q}_v(t) = \frac{\Delta_{st}\omega_v^2}{2(1-S^2)} \left\{A_1 \cos\omega_v t + A_2 \cos\omega_d t + A_3 \cos\left(\omega_b - \frac{\omega_d}{2}\right)t + A_4 \cos\left(\omega_b + \frac{\omega_d}{2}\right)t\right\} \tag{2}$$

where: S is the speed parameter and defined as $S = \frac{\omega_d}{2\omega_b}$.

μ is the frequency ration $\mu = \frac{\omega_d}{2\omega_v}$.

A_1, A_2, A_3, and A_4 are functions of speed parameters and frequency ratio.

ω_d: driving frequency $\omega_d = \frac{2\pi\vartheta}{L}$

ω_v: vehicle frequency $\omega_v = \sqrt{\frac{k_v}{m_v}}$

ω_b: bridge frequency $\omega_b = \frac{\pi^2}{L^2}\sqrt{\frac{EI}{m}}$

Δ_{st}: is approximately the static deflection at mid-span of the beam under the gravity action of the mass m_v at the same point.

Equation (2) shows that the vehicle response is dominated by four frequencies: the driving frequency ω_d, the vehicle frequency ω_v and two shifted frequencies of the beam, $\omega_b - \frac{\omega_d}{2}$ and $\omega_b + \frac{\omega_d}{2}$. By measuring the dynamic response on passing vehicle, all of this frequency can be identified.

To represent the damaged beam, Liu et al. [13] add a mass m_s at the mid-span and derive the formulation in the frequency domain without ignoring non-linear dynamic terms. Then, the mid-span acceleration of the bridge and the acceleration of the vehicle in the frequency domain can be obtained as:

$$\ddot{q}_b\left(\frac{L}{2}, \omega\right) = \sum_{n=1}^{\infty} \frac{2\pi m_v L^3 \omega^2}{i\pi^4 EI - i\omega^2 L^3\left((\rho AL + 2m_s \sin^2\left(\frac{n\pi}{2}\right)\right)} \sin\left(\frac{n\pi}{2}\right) \left\{\ddot{q}_v\left(\omega - \frac{n\pi\vartheta}{L}\right) - \ddot{q}_v\left(\omega + \frac{n\pi\vartheta}{L}\right) + \sqrt{2\pi}\, g\left[\delta\left(\omega - \frac{n\pi\vartheta}{L}\right) - \delta\left(\omega + \frac{n\pi\vartheta}{L}\right)\right]\right\} \tag{3}$$

$$\begin{aligned}\ddot{q}_v(\omega) &= \ddot{q}_v\left(\omega - \frac{2n\pi\vartheta}{L}\right) \\ &- \frac{i\pi^4 EI - i\left(\omega - \frac{n\pi\vartheta}{L}\right)^2 L^3\left((\rho AL + 2m_s \sin^2\left(\frac{n\pi}{2}\right)\right)}{2\pi m_v L^3\left(\omega - \frac{n\pi\vartheta}{L}\right)^2} \ddot{q}_b\left(\frac{L}{2}, \omega - \frac{n\pi\vartheta}{L}\right) \\ &- \sqrt{2\pi}\, g\left[\delta\left(\omega - \frac{2n\pi\vartheta}{L}\right) - \delta(\omega)\right]\end{aligned} \tag{4}$$

Equation (4) indicates that the acceleration of the bridge at $\omega - \frac{n\pi\vartheta}{L}$ and $\omega - \frac{2n\pi\vartheta}{L}$ influence the vehicle response at frequency ω. And not only the first mode frequency but also the high frequency mode contains information of the system and the damage severity level.

3 Experiments and Results

3.1 Lab-Scale Experimental Setup

The experimental vehicle and bridge are shown in Fig. 2. A steel plate supported on both ends was used to model the simply supported bridge. The plate is 2.5 m long and the distance between the two supported is 2 m. The bridge width is 0.35 m and depth is 0.01 m.

Two steel plates supported by a four-point suspension are used to construct the vehicle. The vehicle weighs 5 kg whereas the bridge weighs 80 kg. Four accelerometers are attached on the top plate of the vehicle at the location of four wheels axis. Ten accelerometers are attached to the bridge. Therefore, the bridge's natural frequencies could be measured using both direct and indirect methods. For representing damage, we add mass on the bridge.

The vehicle takes 6 s to across the bridge. The length of the bridge is 2 m, therefore the driving frequency is $\omega_d = 1.04$ Hz.

3.2 Vehicle Measurement

An impact is applied on the top plate of the vehicle to measure the vibration response of the vehicle. To identify the natural frequencies of the vehicle from the acceleration response data of the vehicle, Covariance based Stochastic System Identification (SSI-COV) was used. Figure 3 shows stable poles extracted from the stabilization diagram. The vehicle frequency $\omega_v = 11.21$ Hz was identified in the interval from

Fig. 2 Experimental set up

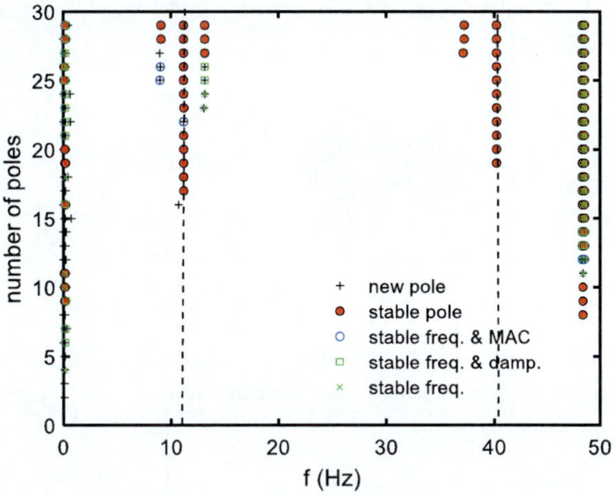

Fig. 3 Stabilization diagram in the interval from 0–50 Hz of passing vehicle

0–50 Hz. This frequency can be considered as the fundamental frequency of the passing vehicle.

3.3 Intact Bridge

Figure 4 shows the vehicle and bridge response, when a vehicle passing through the intact bridge. For the indirect method, the signal stops soon, while for the direct method, the bridge still vibrates after the vehicle exit the bridge.

Four signals from attached accelerometers in the vehicle were analysed using the SSI algorithm. Figure 5 presents the analysed results. Stable poles could be found at three frequencies: $\omega = 6.71\ Hz$, 11.23 Hz, and$\omega = 20.08$ Hz. As pointed in Fig. 3, the second frequency $\omega = 11.23$ Hz is the natural frequency of the vehicle; therefore, the two other frequencies are the ones that are dominated by the bridge. These should be the first and second natural frequencies of the bridge.

For direct measurements, accelerometers were attached directly to the bridge. The analysed results show in Fig. 6. In the frequency interval 0–50 Hz, two bridge natural frequencies can be found $\omega_b = 5.70$ Hz, 21.69 Hz.

3.4 Damaged Bridge

Mass weighted 10kg was added to the bridge to simulate the damaged bridge and to change the bridge natural frequencies. Accelerations recorded at accelerometers No 1, 2 on the vehicle and No 9,10 on the bridge are shown in Fig. 7.

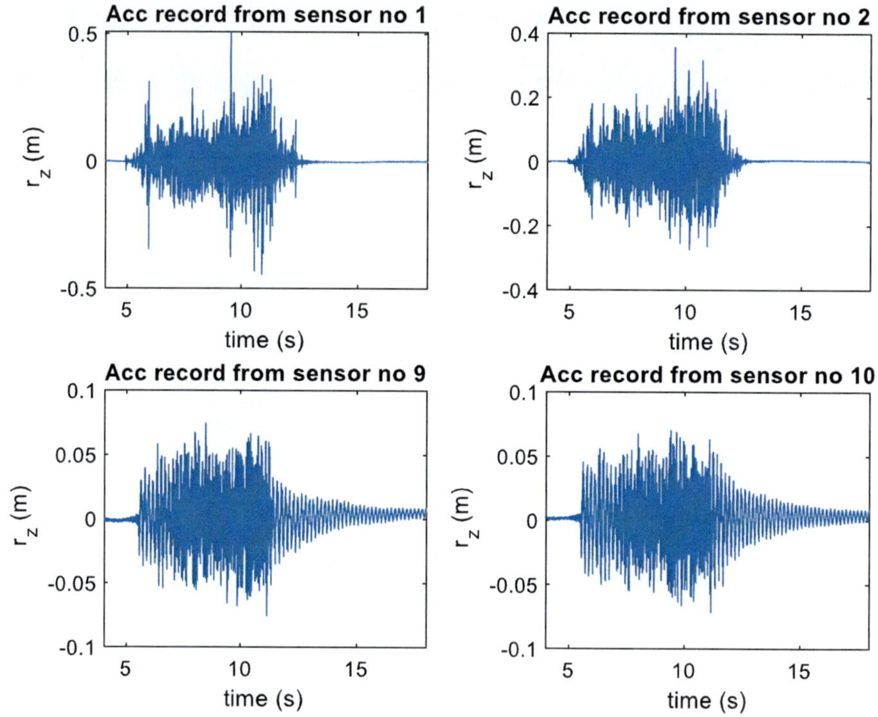

Fig. 4 Acceleration record from sensor no 1, 2 in the vehicle and sensor no 9, 10 in the bridge

Fig. 5 Stabilization diagram in the interval from 0–50 Hz of intact bridge for the indirect measurement

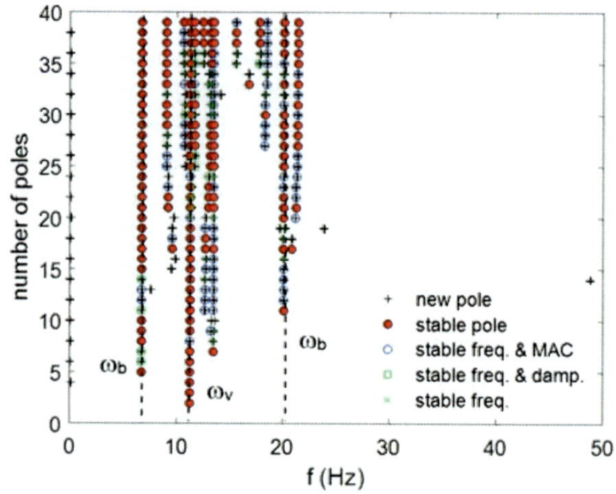

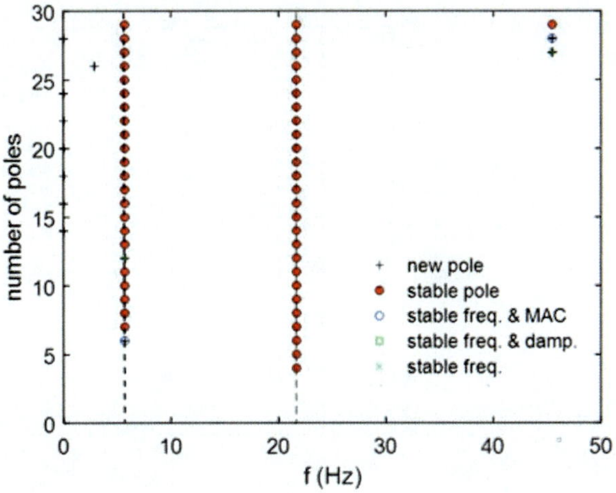

Fig. 6 Stabilization diagram in the interval from 0–50 Hz of intact bridge for the direct measurement

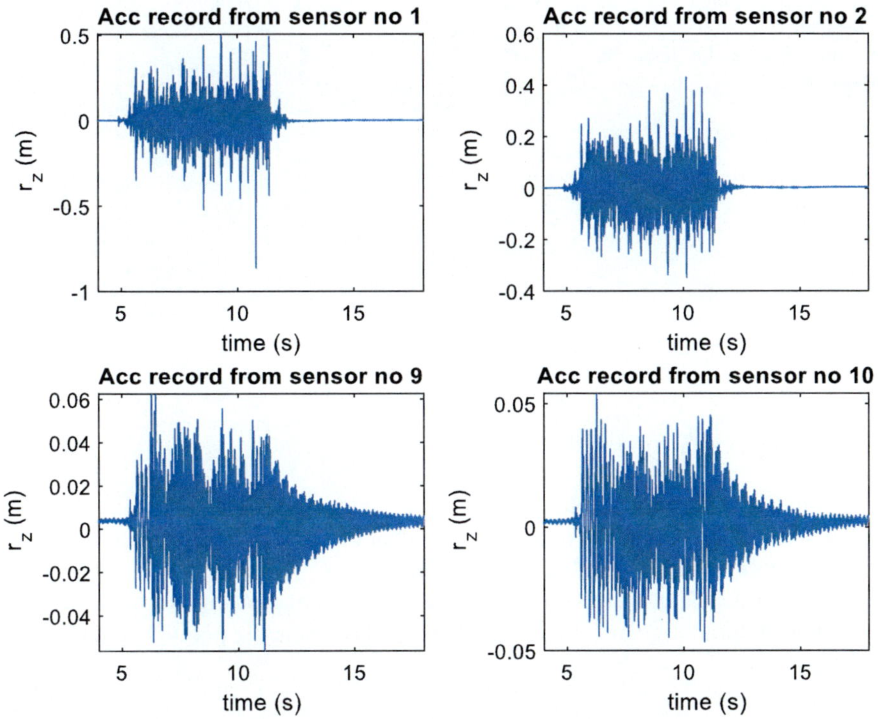

Fig. 7 Acceleration record from sensor no 1, 2 in the vehicle and sensor no 9, 10 in the damaged bridge

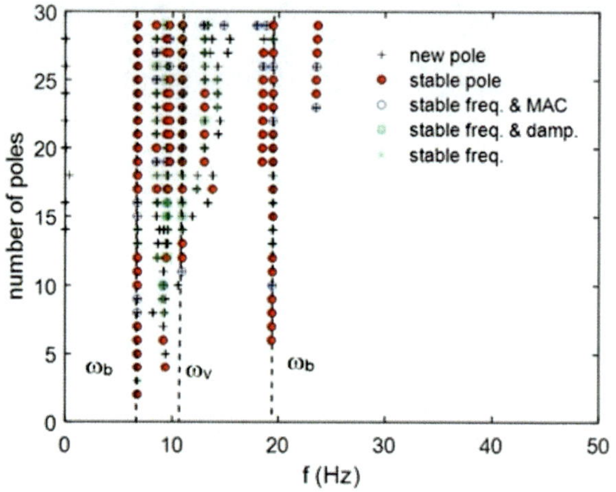

Fig. 8 Stabilization diagram in the interval from 0–50 Hz of damaged bridge for indirect measurement

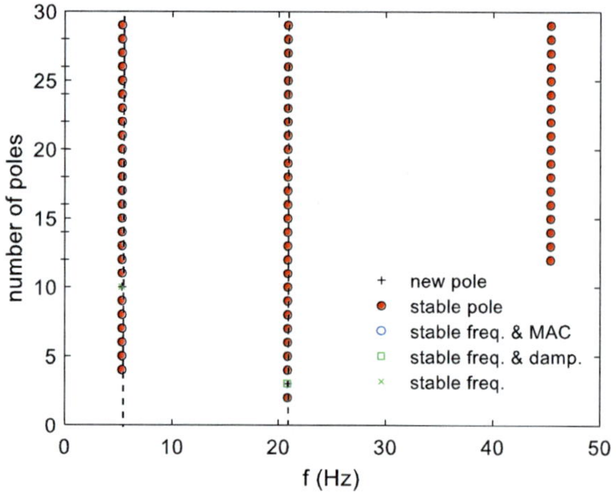

Fig. 9 Stabilization diagram in the interval from 0–50 Hz of damaged bridge for direct measurement

Figure 8 shows some stable poles at three frequencies: $\omega = 6.73$ Hz, 11.3 Hz, and $\omega = 19.4$ Hz. As for the intact case, two natural frequencies of the bridge can be found in the interval from 0–50 Hz of frequency.

Analysing acceleration record from accelerometers attached directly to the bridge, two bridge natural frequencies $\omega_b = 5.31$ Hz, 20.86 Hz clearly pointed out in Fig. 9.

3.5 Discussion

For damaged case, the direct method can clearly show the differences between an intact and damaged bridge is 6,84% and 3.83% for mode 1 and mode 2, respectively. Whereas, for the indirect method, the difference is only 0.3% for mode 1. Therefore,

Table 1 Bridge frequency

Method	Intact		Damage		Differences (%)	
	1st ω_b	2nd ω_b	1st ω_b	2nd ω_b	1st ω_b	2nd ω_b
FEM	5.69	21.64	5.29	20.69		
Direct Method	5.70 (0.18%)	21.69 (0.23%)	5.31 (0.38%)	20.86 (0.82%)	6.84	3.83
Indirect Method	6.71 (17.93%)	20.08 (7.21%)	6.73 (27.22%)	19.40 (6.23%)	0.30	3.39
Differences (%) (between direct and indirect method)	15.05	8.02	21.10	7.53		

using only the vibration response from a passing vehicle is hard to conclude the bridge is damaged or not based on the changes of natural frequencies.

Table 1 summarizes all frequencies from the finite element model in Sap2000, direct, and indirect methods of the intact and damaged bridge, respectively. The frequencies got from the direct method are more accurate when compared with the FEM results. The indirect method cannot measure the absolute values of the natural bridge frequencies. That means there is a difference between the frequencies got from the direct and indirect methods. This difference in mode 1 is bigger than in mode 2. This is understandable because as discussed above, the frequency indirect method is dominated by the frequency band between $\omega_b - \frac{\omega_d}{2}$ and $\omega_b + \frac{\omega_d}{2}$.

For damaged case, the direct method can clearly show the differences between an intact and damaged bridge is 6,84% and 3.83% for mode 1 and mode 2, respectively. Whereas, for the indirect method, the difference is only 0.3% for mode 1. Therefore, using only the changes of natural frequencies extracted from a passing vehicle is hard to conclude the bridge is damaged or not.

4 Conclusion

In this paper, a vehicle with strung mass was fabricated to measure the vibration response of a laboratory bridge. The laboratory bridge with the simply supported boundary condition was made from steel plates. The direct method is that accelerometers are attached directly to the bridge surface and the indirect method is that accelerometers are attached to the passing vehicle. Both methods were used in this research. Damage was introduced on the bridge by adding mass. Comparing the natural frequencies got from the two methods, the conclusion is that the indirect method can be used to measure the vibration response of the bridge. However, the indirect method only can show the bridge natural frequencies between a frequency band. This frequency band depends on the natural frequencies of the bridge and the speed of the vehicle. Moreover, the indirect method shows difficulties to detect damage due to small changes in the natural frequencies.

References

1. Nguyen DH, Nguyen QB, Bui-Tien T, De Roeck G, Wahab MA (2020) Damage detection in girder bridges using modal curvatures gapped smoothing method and convolutional neural network: application to Bo Nghi bridge. Theor Appl Fract Mech 109:102728
2. Nguyen DH, Bui TT, De Roeck G, Wahab MA (2019) Damage detection in Ca-Non Bridge using transmissibility and artificial neural networks. Struct Eng Mech 71(2):175–183
3. Mazurek DF, DeWolf JT (1990) Experimental study of bridge monitoring technique. J Struct Eng 116(9):2532–2549
4. McLamore VR, Hart GC, Stubbs IR (1971) Ambient vibration of two suspension bridges. J Struct Div 97(10):2567–2582
5. Yang Y-B, Lin C, Yau J (2004) Extracting bridge frequencies from the dynamic response of a passing vehicle. J Sound Vib 272(3–5):471–493
6. Lin C, Yang Y (2005) Use of a passing vehicle to scan the fundamental bridge frequencies: an experimental verification. Eng Struct 27(13):1865–1878
7. O'Brien EJ, Malekjafarian A (2015) Identification of bridge mode shapes using a passing vehicle. In: 7th international conference on structural health monitoring of intelligent infrastructure, Torino, Italy, July, 2015. International Society for Structural Health Monitoring of Intelligent
8. Siringoringo DM, Fujino Y (2012) Estimating bridge fundamental frequency from vibration response of instrumented passing vehicle: analytical and experimental study. Adv Struct Eng 15(3):417–433
9. Yang Y, Chang K, Li Y (2013) Filtering techniques for extracting bridge frequencies from a test vehicle moving over the bridge. Eng Struct 48:353–362
10. Sitton JD, Zeinali Y, Rajan D, Story BA (2020) Frequency estimation on two-span continuous bridges using dynamic responses of passing vehicles. J Eng Mech 146(1):04019115
11. Zhang Y, Wang L, Xiang Z (2012) Damage detection by mode shape squares extracted from a passing vehicle. J Sound Vib 331(2):291–307
12. Yang Y, Yang JP (2018) State-of-the-art review on modal identification and damage detection of bridges by moving test vehicles. Int J Struct Stab Dyn 18(02):1850025
13. Liu J, Chen S, Bergés M, Bielak J, Garrett JH, Kovačević J, Noh HY (2020) Diagnosis algorithms for indirect structural health monitoring of a bridge model via dimensionality reduction. Mech Syst Signal Process **136**:106454

Topology Optimization for a Large-Scale Truss Bridge Using a Hybrid Metaheuristic Search Algorithm

H. Tran-Ngoc, H. Nguyen-Manh, H. Viet Tran, Q. Nguyen-Huu, N. Hoang-Thanh, T. Le-Xuan, T. Bui-Tien, N. Nguyen-Cam, and M. Abdel Wahab

Abstract This paper introduces a simple and effective method to optimize the weight and geometry of truss members of a large-scale truss bridge. Design variables comprise the redundant forces acting on the structure and the cross-sectional dimensions of truss members. The objective function is constraints on member stresses chosen to minimize the deviation between calculated and desired results. To increase global search capabilities, a hybrid metaheuristic algorithm (HGAPSO) combining genetic algorithm (GA) and particle swarm optimization (PSO) is employed to look for the best solution. The results show that HGAPSO not only provides the considered structure with a lighter weight than the original design but also outperforms the traditional optimization algorithm (GA, PSO) in terms of the search capacity for the most optimal solution.

Keywords Large-scale truss bridge · Topology optimization · GA · PSO · A hybrid metaheuristic algorithm

---

H. Tran-Ngoc (✉) · M. A. Wahab
Department of Electrical Energy, Metals, Mechanical Constructions, and Systems, Faculty of Engineering and Architecture, Ghent University, 9000 Gent, Belgium
e-mail: hoa.tran@ugent.be

H. Tran-Ngoc · H. Nguyen-Manh · Q. Nguyen-Huu · T. Le-Xuan · T. Bui-Tien · N. Nguyen-Cam
Department of Bridge and Tunnel Engineering, Faculty of Civil Engineering, University of Transport and Communications, Hanoi, Vietnam

H. V. Tran
Department of Structural Mechanics, Faculty of Civil Engineering, University of Transport and Communications, Hanoi, Vietnam

N. Hoang-Thanh
Department of Science and Technology, The Ministry of Transport, Hanoi, Vietnam

M. Abdel Wahab (ed.), *Proceedings of the 2nd International Conference on Structural Damage Modelling and Assessment*, Lecture Notes in Civil Engineering 204,
https://doi.org/10.1007/978-981-16-7216-3_4

1 Introduction

In recent decades, optimization has been received special attention from the scientific community [1–11]. In many fields of optimization, topology optimization is a mathematical method that optimizes geometry, material, boundary conditions, constraints, etc., all of which play a vital role in maximizing the performance of the system. Martin et al. [12] proposed a novel topology optimization approach for tall buildings under seismic loading utilizing the modal decomposition method. The obtained results concluded that the soil and seismological effects beneath the structure created considerable impacts on topology optimization for the considered structures. Sergeyev and Mroz [13] applied sensitivity analysis techniques for the optimal design of 3D frame structures. Joint positions and cross-sectional parameters were selected as parameters that needed to optimize, in which free frequency constraints and imposed stress were considered as an objective function. Zhao et al. [14] introduced a new pseudo-excitation-based method for structural topology optimization. In their work, the effect of stationary random excitations on the structure was assessed. The results showed that the proposed method possibly increased accuracy and reduced computational costs.

Over the recent decades, optimization algorithms, especially, particle swarm optimization (PSO) have been successfully applied in many fields [15–18]. Tran et al. [19] applied genetic algorithm (GA) and PSO to identify structural dynamic behavior and determine uncertain parameters including material properties and boundary conditions for a large-scale truss bridge. Khatir et al. [20] employed PSO combined with an Artificial neural network (ANN) to detect damages in steel structures based on the objective function of natural frequencies. Tran et al. [21] updated uncertain parameters for a multi-span cable-stayed bridge using a hybrid algorithm of PSO.

However, PSO has fundamental disadvantages that can reduce its effectiveness in solving optimization problems. A major drawback of PSO is that this algorithm not only relies heavily on the quality of the original population but also lacks capabilities such as crossover or mutation operators to improve the quality of the new generations. If the positions of the initial populations are far from the global best, it is difficult to find the best solution.

Therefore, in this work, we propose combining PSO with GA (HGAPSO) to deal with structural topology optimization of a large-scale truss bridge. The proposed algorithm can combine the advantages of GA (capacity of crossover and mutation to improve the quality of new populations) and PSO (capability of stochastic global search). The proposed algorithm can overcome the disadvantages of traditional GA, and PSO separately, improving the accuracy of the obtained results.

2 HGAPSO

The working principle of HGAPSO is presented in detail as below steps:

Step 1. Generating initial populations with features: position (P_0), velocity (V_0), the local best (L_0), and the global best (G_0), respectively.

$$P_0 = \left[p_0^1, p_0^2, \ldots, p_0^i\right]; i = [1; n] \tag{1}$$

$$V_0 = \left[v_0^1, v_0^2, \ldots, v_0^i\right] \tag{2}$$

$$L_0 = \left[l_0^1, l_0^2, \ldots, l_0^i\right]; G_0 \tag{3}$$

$$p_0^i = \left[p_0^{1i}, p_0^{2i}, \ldots, p_0^{mi}\right] \tag{4}$$

$$v_0^i = \left[v_0^{1i}, v_0^{2i}, \ldots, v_0^{mi}\right] \tag{5}$$

$$l_0 = \left[l_0^{1i}, l_0^{2i}, \ldots, l_0^{mi}\right] \tag{6}$$

n indicates the number of uncertain parameters that need to be found; m is the number of populations.

Step 2. Select the objective function $W(P)$

$$W(P) = min(W) \tag{7}$$

W is the weight of the bridge.

Step 3. Apply the constraints

$$\sigma_c^{t(q)} < \sigma_p^t \tag{8}$$

$$\sigma_c^{c(q)} < \sigma_p^c \tag{9}$$

$\sigma_c^{t(q)}, \sigma_c^{c(q)}$ are calculated tension stress, and calculated compression stress of truss member q^{th}, respectively. $q \in [1, K]$. K is the total truss member. σ_p^t andσ_p^c are allowance tension stress, and allowance compression stress, respectively. σ_c^q andσ_p^c are referred from bridge design standard TCN19-79.σ_c^q and$\sigma_p^c = 600{,}000\ kN/m^2$.

Step 4. The local best of particles is determined and put in an increasing order based on the objective function $\varnothing(P)$

$$L_0 = \left[L_0^{max} \ldots L_0^{min}\right] \tag{10}$$

Step 5. The parents with the best quality determined from step 2 are selected for crossover and mutation, the number of parents selected based on the percentage of crossover and mutation is calculated in Eqs. 11 and 12.

$$n_c = p_c * n \tag{11}$$

$$n_m = p_m * n \tag{12}$$

p_c is crossover percent and p_m is mutation percent.

Step 6. Repeat the process from step 2 to step 4 until termination criteria are satisfied.

Step 7. Using particles obtained from step 5 to look for the best solution based on the global search capacity of PSO.

Step 7.1. Updated velocity and position of particles.

$$V_{t+1}^{j} = w * V_t^j + c_1 * r_1 * \left(L_t^{best} - P_t^j\right) + c_2 * r_2 * \left(G_t^{best} - P_t^j\right) \tag{13}$$

$$P_{t+1}^{j} = \left(P_t^j + V_{t+1}^j\right);\ ;\ j = [1, m] \tag{14}$$

$$If(P_{t+1}^j > P_{max}) \tag{15}$$

$$P_{t+1}^j = P_{min} + rand * P_{min} \tag{16}$$

$$If(P_{t+1}^j < P_{min}) \tag{17}$$

$$P_{t+1}^j = P_{min} + rand * P_{min} \tag{18}$$

Where P_t^j, P_{t+1}^j are the positions, V_t^j, V_{t+1}^j indicate the velocitys of element j at time t and t+1, respectively. c_1 and c_2 are the learning factor, whereas r_1 and r_2 denote random numbers (0 < $r_1$andr_2< 1). While w is the inertia weight parameter, L_t^{best} and G_t^{best} indicate the local best, and the global best at the time t, respectively. P_{min} and P_{max} are lower and upper bound of the search area.

Step 7.2. Select the local best of each element, and the global best for the next iteration based on the objective function $W(P)$.

$$If(W\left(P_{t+1}^j\right) < W\left(P_t^j\right)) \tag{19}$$

$$W(L_{t+1}^{best}) = W\left(P_{t+1}^j\right);\ L_{t+1}^{best} = P_{t+1}^j \tag{20}$$

$$Otherwise\left(W\left(L_{t+1}^{best}\right) = W\left(P_t^j\right); L_{t+1}^{best} = P_t^j\right) \quad (21)$$

$$W(G_{t+1}^{best}) = \min(W\left(L_{t+1}^{best}\right)) \quad (22)$$

Step 8. Repeat step 6 until termination criteria are satisfied.
Step 9. The iteration complete and the best solution is obtained.

$$f(G^{best}, k_0) = \min(W(\text{P})) \quad (23)$$

$$G^{best} = P(k_0) \quad (24)$$

k_0 is ${k_0}^{th}$ iteration, $k_0 \in [0, \text{N}]$. N is the total number of iteration.

3 Topology Optimization for a Large-Scale Truss Bridge

3.1 Bridge Description

To assess the effectiveness of HGAPSO, in this section, a large-scale truss bridge (Guadalquivir bridge in Seville, Spain) is employed for structural topology optimization. The bridge consists of five spans with the same length for each span (50 m). The layout of the bridge is depicted in Fig. 1.

3.2 Finite Element Model (FEM)

The FEM of the bridge (Fig. 2) is built using MATLAB toolbox Stabil [22]. The bridge consists of 564 nodes, 1643 elements, and 8 section types of truss members

Fig. 1 The Guadalquivir railway bridge

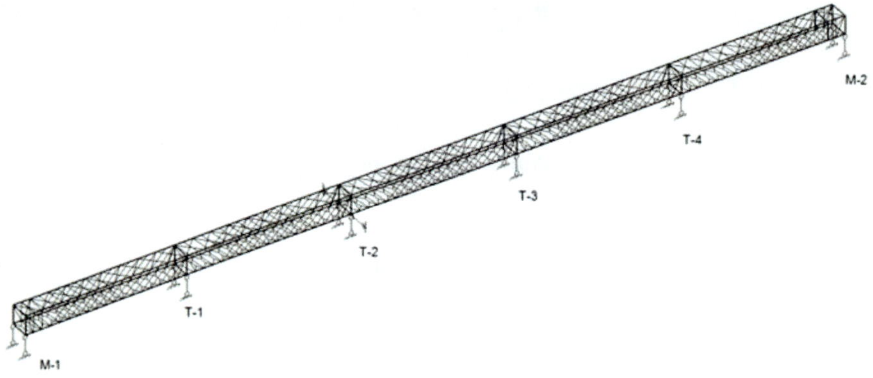

Fig. 2 FEM of the bridge

Table 1 Material properties

Components	Values	Unit
Young's modulus	2.15×10^{11}	N/m^2
Weight density	7850	Kg/m^3
Poisson's ratio	0.3	

(upper chords, vertical chords, diagonal chords, lower chords, cross girders, top chord bracing truss members, bottom chord bracing truss members, and stringers as shown in Table 2. Those members are modelled by utilizing three-dimensional elements. Each element has three rotational displacements and three translational displacements. Other structures comprising maintenance path, power line, handrail, and rail track, are assessed as an added mass in the model.

It is assumed that all the joint nodes at the lower chords are subjected to a concentrated force F = 100 kN. The material properties of steel truss are listed as Table 1.

The geometry of truss members is described in Table 2.

The calculated maximum stress of the truss members is shown in Table 3.

The total steel weight of the bridge for one span is: 247.15 (tons).

3.3 *Topology Optimization and Results*

In this section, the proposed approach (HGAPSO) is used to optimize the dimensions of truss members and the weight of the bridge. In order to compare with HGAPSO, GA and PSO are also employed. The population is 100 for all algorithms. Crossover and mutation operators of GA and HGAPSO are 0.1 and 0.8, respectively. For PSO and, HGAPSO, the inertia weight parameter (w) is 0.3, and the values of the cognition learning factor and the social learning factor are $c_1 = 2$ and $c_2 = 2$. The objective

Table 2 Geometric properties of truss members

Components	Properties	Values/Unit
Upper chords	(Upper, lower) flange width	0.3/m
	(Upper, lower) flange thickness	0.016/m
	Depth between	0.6/m
	Web thickness	0.02/m
	Area	0.0216/m^2
	Moment of main inertia	3.743^{-04}/m^4
Lower chords	(Upper, lower) flange width	0.3/m
	(Upper, lower) flange thickness	0.016/m
	Depth between	0.6/m
	Web thickness	0.02/m
	Area	0.0216/m^2
	Moment of main inertia	3.743^{-04} m^4
Vertical chords	(Upper, lower) flange width	0.3/ m
	(Upper, lower) flange thickness	0.016/m
	Depth between	0.6/m
	Web thickness	0.02/m
	Area	0.0216/m^2
	Moment of main inertia	3.743^{-04}/m^4
Cross girder	(Upper, lower) flange width	0.3/m
	(Upper, lower) flange thickness	0.016/m
	Depth between	0.6/m
	Web thickness	0.02/m
	Area	0.0216/m^2
	Moment of main inertia	3.743^{-04}/m^4
Diagonal chords	(Upper, lower) flange width	0.3/m
	(Upper, lower) flange thickness	0.016/m
	Depth between	0.6/m

(continued)

function is the weight of the bridge. The stop criteria of loops of all algorithms are established as follows: the weight of the bridge calculated by PSO is lower than that of the initial design or the maximum number of iterations is 300.

Table 4 shows the geometric properties of truss members after optimization using HGAPSO.

From Table 5, after optimization, the total steel weight of the bridge reduces significantly. The total steel weight calculated by HGAPSO is the lowest, at, 202.78 tons, whereas those calculated by PSO and GA are 218.08 tons and 231.78 tons, respectively.

Table 2 (continued)

Components	Properties	Values/Unit
	Web thickness	0.02/m
	Area	0.0216/m^2
	Moment of main inertia	3.743^{-04}/m^4
Top chord bracing	(Upper, lower) flange width	0.1/m
	(Upper, lower) flange thickness	0.01/m
	Depth between	0.2/m
	Web thickness	0.01/m
	Area	0.004/m^2
	Moment of main inertia	6.893^{-06}/m^4
Bottom chord bracing	(Upper, lower) flange width	0.1/m
	(Upper, lower) flange thickness	0.01/m
	Depth between	0.2/m
	Web thickness	0.01/m
	Area	0.004/m^2
	Moment of main inertia	6.893^{-06}/m^4
Stringers	(Upper, lower) flange width	0.1/m
	(Upper, lower) flange thickness	0.01/m
	Depth between	0.2/m
	Web thickness	0.01/m
	Area	0.004/m^2
	Moment of main inertia	6.893^{-06}/m^4

Table 3 Calculated maximum stress before optimization

Maximum stress	Value (kN/m^2)
Tension	42,386.30
Compression	52,982.87

4 Conclusion

This paper proposes a simple and effective approach to optimize the dimensions of the truss members and the weight of a large-scale steel truss bridge. To increase the effectiveness of global search capabilities, HGAPSO is used. HGAPSO possibly the gain advantages of both GA (capacity of crossover and mutation to improve the quality of new populations) and PSO (capability of stochastic global search). From obtained results, some main conclusions can be drawn.

–After optimization, the total steel weight of the bridge reduces significantly, (a reduction of approximately 18% compared with the initial design).

Table 4 Geometric properties of truss members after optimization using HGAPSO

Components	Properties	Values/Unit
Upper chords	(Upper, lower) flange width	0.2892/m
	(Upper, lower) flange thickness	0.016/m
	Depth between	0.5834/m
	Web thickness	0.02/m
	Area	$2.09^{-02}/m^2$
	Moment of main inertia	$3.43^{-04}/m^4$
Lower chords	(Upper, lower) flange width	0.4556/m
	(Upper, lower) flange thickness	0.016/m
	Depth between	0.5435/m
	Web thickness	0.02/m
	Area	$2.545^{-02}/m^2$
	Moment of main inertia	$3.609^{-04}/m^4$
Vertical chords	(Upper, lower) flange width	0.1645/m
	(Upper, lower) flange thickness	0.016/m
	Depth between	0.1814 /m
	Web thickness	0.02/m
	Area	$0.889^{-02}/m^2$
	Moment of main inertia	$0.114^{-04}/m^4$
Cross girder	(Upper, lower) flange width	0.2496/m
	(Upper, lower) flange thickness	0.016/m
	Depth between	0.5111/m
	Web thickness	0.02/m
	Area	$1.82^{-02}/m^2$
	Moment of main inertia	2.311^{-04}
Diagonal chords	(Upper, lower) flange width	0.2496/m
	(Upper, lower) flange thickness	0.016/m
	Depth between	0.4003 /m

(continued)

–HGAPSO provides a better solution for topology optimization of the considered structure than GA and PSO.
–In this paper, only the dimensions of truss members and the weight of the bridge are taken into account for optimization. Further studies will be carried out to optimize the geometry, connection for truss bridges as well as other structures.

Table 4 (continued)

Components	Properties	Values/Unit
	Web thickness	0.02/m
	Area	1.466^{-02}/m^2
	Moment of main inertia	1.117^{-04}/m^4
Top chord bracing	(Upper, lower) flange width	1.116^{-04}/m
	(Upper, lower) flange thickness	0.01/m
	Depth between	1.1167^{-04}/m
	Web thickness	0.01/m
	Area	0.3328^{-02}
	Moment of main inertia	4.3915^{-06}/m^4
Bottom chord bracing	(Upper, lower) flange width	0.0775/m
	(Upper, lower) flange thickness	0.01/m
	Depth between	0.1662/m
	Web thickness	0.01/m
	Area	0.321^{-02}/m^2
	Moment of main inertia	3.942^{-06}/m^4
Stringers	(Upper, lower) flange width	0.1/m
	(Upper, lower) flange thickness	0.01/m
	Depth between	0.2/m
	Web thickness	0.01/m
	Area	0.004/m^2
	Moment of main inertia	6.893^{-06}/m^4

Table 5 The weight of the bridge (one span) after optimization

Algorithms	Weight (tons) for one span
Design	247.15
GA	231.78 (6.22%)
PSO	218.08 (11.76%)
HGAPSO	202.78 (17.95%)

Acknowledgements The authors acknowledge the financial support of BOF of Ghent University, the project research "B2021-GHA-04" of the Ministry of Education and Training

References

1. Tran-Ngoc H, Khatir S, Le-Xuan T, De Roeck G, Bui-Tien T, Wahab MA (2021) Finite element model updating of a multispan bridge with a hybrid metaheuristic search algorithm using experimental data from wireless triaxial sensors. Eng Comput 1–19
2. Ho LV, Nguyen DH, Mousavi M, De Roeck G, Bui-Tien T, Gandomi AH, Wahab MA (2021) A hybrid computational intelligence approach for structural damage detection using marine predator algorithm and feedforward neural networks. Comput Struct 252:106568
3. Dang HV, Raza M, Tran-Ngoc H, Bui-Tien T, Nguyen HX (2021) Connection stiffness reduction analysis in steel bridge via deep CNN and modal experimental data. Struct Eng Mech 77(4):495–508
4. Tran-Ngoc H, Khatir S, De Roeck G, Bui-Tien T, Wahab MA (2019) An efficient artificial neural network for damage detection in bridges and beam-like structures by improving training parameters using cuckoo search algorithm. Eng Struct 199:109637
5. Khatir S, Boutchicha D, Le Thanh C, Tran-Ngoc H, Nguyen TN, Abdel-Wahab M (2020) Improved ANN technique combined with Jaya algorithm for crack identification in plates using XIGA and experimental analysis. Theor Appl Fract Mech 107:102554
6. Tran-Ngoc H, Khatir S, De Roeck G, Bui-Tien T, Wahab MA (2020) Damage assessment in beam-like structures using cuckoo search algorithm and experimentally measured data. In: Proceedings of the 13th international conference on damage assessment of structures. Springer, Singapore, pp 380–385
7. Dang HV, Tran-Ngoc H, Nguyen TV, Bui-Tien T, De Roeck G, Nguyen HX (2020) Data-driven structural health monitoring using feature fusion and hybrid deep learning. IEEE Trans Autom Sci Eng
8. Tran-Ngoc H, Khatir S, Ho-Khac H, De Roeck G, Bui-Tien T, Wahab MA (2021) Efficient Artificial neural networks based on a hybrid metaheuristic optimization algorithm for damage detection in laminated composite structures. Compos Struct 262:113339
9. Nguyen DH, Tran-Ngoc H, Bui-Tien T, De Roeck G, Wahab MA (2020) Damage detection in truss bridges using transmissibility and machine learning algorithm: Application to Nam O bridge. Smart Struct Syst 26(1):35–47
10. Ho VL, Tran NH, De Roeck G, Bui TT, Wahab MA (2018) System identification based on vibration testing of a steel I-beam. In: International conference on numerical modelling in engineering. Springer, Singapore, pp 254–268
11. Ho LV, Khatir S, Roeck GD, Bui-Tien T, Wahab MA (2020) Finite element model updating of a cable-stayed bridge using metaheuristic algorithms combined with Morris method for sensitivity analysis. Smart Struct Syst 26(4):451–468
12. Martin A, Deierlein GG (2020) Structural topology optimization of tall buildings for dynamic seismic excitation using modal decomposition. Eng Struct 216:110717
13. Sergeyev O, Mroz Z (2000) Sensitivity analysis and optimal design of 3D frame structures for stress and frequency constraints. Comput Struct 75(2):167–185
14. Zhao X, Wu B, Lai SK, Li Z, Zhong H (2021) A PEM-based topology optimization for structures subjected to stationary random excitations. Eng Struct 229:111613
15. Tran-Ngoc H, Khatir S, Le-Xuan T, De Roeck G, Bui-Tien T, Wahab MA (2020) A novel machine-learning based on the global search techniques using vectorized data for damage detection in structures. Int J Eng Sci 157:103376
16. Tran-Ngoc H, He L, Reynders E, Khatir S, Le-Xuan T, De Roeck G, ... Wahab MA (2020) An efficient approach to model updating for a multispan railway bridge using orthogonal diagonalization combined with improved particle swarm optimization. J Sound Vib 476:115315
17. Tran NH, Bui TT (2019) Damage detection in a steel beam structure using particle swarm optimization and experimentally measured results. Sci J Transport 9:3–9
18. Nguyen-Ngoc L, Tran H, Bui-Tien T, Mai-Duc A, Abdel Wahab M, Nguyen HX, , De Roeck G (2021) Damage detection in structures using particle swarm optimization combined with artificial neural network. Smart Struct Syst 28(1):1–12

19. Tran-Ngoc H, Khatir S, De Roeck G, Bui-Tien T, Nguyen-Ngoc L, Abdel Wahab M (2018) Model updating for Nam O bridge using particle swarm optimization algorithm and genetic algorithm. Sensors 18(12):4131
20. Khatir S, Khatir T, Boutchicha D, Le Thanh C, Tran H, Bui TQ, …, Abdel Wahab M (2020) An efficient hybrid TLBO-PSO-ANN for fast damage identification in steel beam structures using IGA. Smart Struct Syst 25(5):605–617
21. Tran H, Khatir S, De Roeck G, Nguyen L, Bui TT, Abdel Wahab M (2020) An efficient approach for model updating of a large-scale cable-stayed bridge using ambient vibration measurements combined with a hybrid metaheuristic search algorithm. Smart Struct Syst 25(4):487–499
22. Dooms D, Jansen M, De Roeck G, Degrande G, Lombaert G, Schevenels M, François S (2010) StaBIL: a finite element toolbox for MATLAB. VERSION 2.0 USER'S GUIDE

Predicting the Displacement of Diaphragm Wall for Deep Excavation Problem on the Basing Thickly Soft Soil in an Urban Area Using Semi-Top-Down Construction Method

Thanh Sang-To, Minh Hoang-Le, Magd Abdel Wahab, and Thanh Cuong-Le

Abstract In this study, the finite element method (FEM) is employed in the forecasting the displacement of diaphragm wall. A popular and effective constructed method, termed semi-top-down (STD), is adopted to reduce the displacement of the diaphragm wall as well as ground settlement. The response of soil and diaphragm walls is thoroughly investigated via the Hardening Soil (HS) model. An actual project is employed to evaluate this simulation. The obtained results draw a preliminary picture of the estimating displacement of a diaphragm wall and the proper way to forecast soil parameters excavation problems of a high-rise building on base thick, soft soil as the in Ho Chi Minh City, Vietnam.

Keywords F.E.M. · Deep excavation

1 Introduction

These days, a variety of large-scale construction requires deeper excavation. Lack of resistance and excessive deformation are popularly causing problems for failures when too deep excavation is necessary. This leads to stability concerns at final-stages of excavation, even sometimes to complete collapse. Each project must use a proper method to carry out excavation depending on the conditions in the field. Some vital areas are identified by simulations using numerical analysis. Prediction of the lateral displacement of diaphragm wall is a vital section in the design problem of excavation. The excavation process can have an adverse effect on both adjacent buildings and other existing infrastructure. FEM commercial software's, namely ABAQUS [1, 2], PLAXIS [3, 4], and so on, are very essential tool to design and for early warning in Geotechnics, especially in deep excavation problems. FEM provides advanced

---

T. Sang-To (✉) · M. Abdel Wahab
Soete Laboratory, Faculty of Engineering and Architecture, Ghent University, Technologiepark Zwijnaarde 903, B-9052 Zwijnaarde, Belgium
e-mail: sang.tt@ou.edu.vn

T. Sang-To · M. Hoang-Le · T. Cuong-Le
Faculty of Civil Engineering, Ho Chi Minh City Open University, Ho Chi Minh City, Vietnam

M. Abdel Wahab (ed.), *Proceedings of the 2nd International Conference on Structural Damage Modelling and Assessment*, Lecture Notes in Civil Engineering 204,
https://doi.org/10.1007/978-981-16-7216-3_5

models to analyze staged construction of buildings and their impact on adjacent structures effectively and reliably.

In recent years, an outstanding excavation process, called the semi-top-down technique, has been popularly employed for the excavation of several traffic infrastructure constructions, such as subway stations [5–8] and high-rise building's basements [9, 10]. With a STD construction approach, the soils above the middle slab of the substructure are excavated following the bottom-up process; then, the middle slab is constructed. Since the poured middle slab is full stiffness, soil removal continues to the final level. Due to excavation of the upper pit following the bottom-up process, the STD approach spends less time in excavation than the top-down approach. Meanwhile, as the middle slab has been constructed before excavating the lower pit, its supporting system is stiffer than that of the bottom-up technique. Furthermore, in addition to reducing the harmful effect on the adjacent buildings, it shortens the construction duration as much as possible. Based on these advantages, the STD approach in combination with the local bracing system technique at some opening positions of concrete slab was selected to construct a high-rise building.

2 Simulation Using F.E.M

In this study, a high-rise building (Sunrise City View-SCV) is adopted to evaluate the effectiveness of the simulation. With the special location, the SCV project is reasonable to test the simulation with a semi-top-down construction method and advanced HS. model.

2.1 Overview of the Sunrise City View Building

In this paper, excavation for the building with three basements on based thick and soft soil is introduced. This building is located in district 7, Ho Chi Minh city, high constructed density with many high-rise buildings, as shown in Fig. 1. The dimensions of the excavation are 150.9 m and 126.5 m, for the length and width, respectively. This project is excavated to a depth of 16.45 m. Besides, a reinforced concrete diaphragm wall, 0.8 m thick and 26 m deep, is erected as a retaining wall of the excavation. Figure 2 illustrates a typical cross-section of the excavation. The Semi-Top-Down approach is selected for this high-rise building construction. The excavation details are shown in Table 1.

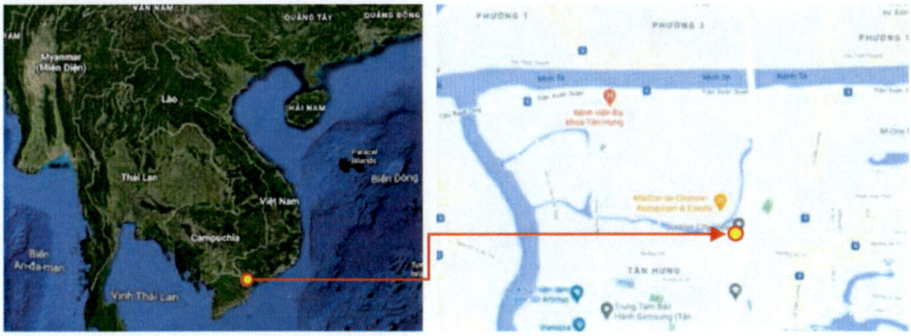

Fig. 1 Location of the Sunrise City View

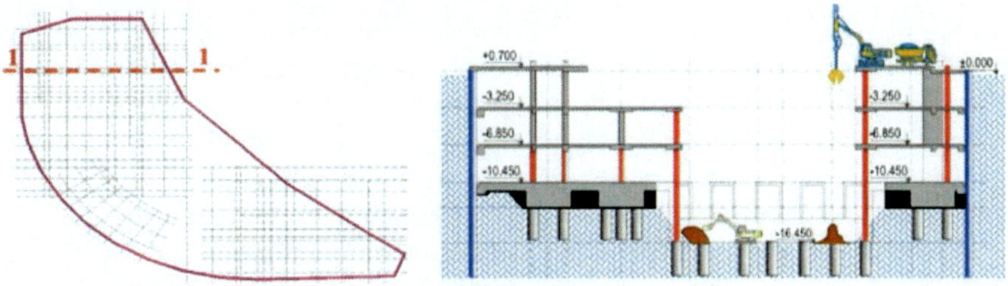

Layout plan diaphragm wall Section 1-1

Fig. 2 Cross-section of the excavation of S.C.V. project

Table.1 Construction sequence of the Sunrise City View project

Stage	Main excavation activity
1	Construct diaphragm wall
2	Excavate to 1.65 m from the ground surface
3	Construct ground floor
4	Excavate to 3.75 m from the ground surface
5	Construct 1st basement slab
6	Excavate to 7.35 m from the ground surface
7	Construct 2nd basement slab
8	Excavate to 12.95 m from the ground surface
9	Construct 3rd basement slab and cap foundation
10	Excavate to 16.45 m from the ground surface

2.2 Soil Model and Parameter

In the study, the hardening soil model is employed to predict the horizontal displacement of the diaphragm wall. The HS model is an advanced model to simulate the

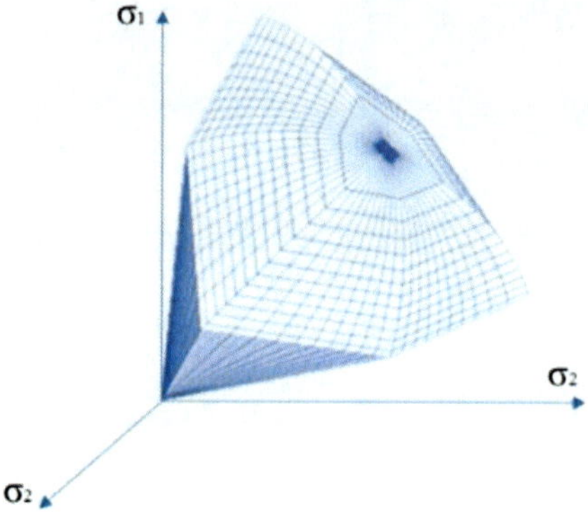

Fig. 3 The H.S. model yield surface in principal stress space (c = 0)

behavior of a variety of soils from soft soils to stiff soils. It is interesting to note that its soil stiffness depends on stress, and it obeys the power law as illustrated in Eq. (1). Additionally, the Hs model also considers plastic straining due to primary deviatoric loading (E_{50}) and primary compression (E_{oed}). The elastic un/reloading (E_{ur} and υ_{ur}), dilatancy effect and failure are following the MohrCoulomb (MC) criterion. To more understand the yield surfaces of the Hardening soil model, reader may refer to Fig. 3. The hexagon is the shape of the yield cap based on the classical MC failure criterion. It describes the yield surfaces according to three-principal stress, including, and the limiting states of stress are represented by means of the friction angle φ', the cohesion c' and the dilatancy angle ψ. Soil stiffnesses are simulated closer to really by defining 3 distinct stiffnesses, namely E_{50}^{ref} : triaxial loading stiffness, E_{ur}^{ref} : triaxial unloading stiffness and E_{oed}^{ref}: oedometer loading stiffness.

$$E_{50} = E_{50}^{ref}\left(\frac{c' \times \cos(\varphi') + \sigma_3^{'} \times \sin(\varphi')}{c' \times \cos(\varphi') + p_{ref} \times \sin(\varphi')}\right)^m \quad (1)$$

where p_{ref} is reference stress (default normally equal to 100 units).

Stratigraphy of SCV with three clay layers, from soft to stiff, up to 50 m is interspersed with a sand layer. The value of belongs to a range of [0, 45] and the density of the soil layer varies from 15 to 19.9 kN/m3. The ground profile and soil parameters of the SCV are summarized in Table 2 and Fig. 4. Field observations provided that the groundwater level was approximately 1.5 m below the surface ground.

The rest of stiffness parameters, the E_{ur}^{ref} and E_{oed}^{ref} values are selected, respectively, to be equal to 3 × E_{50}^{ref} and E_{50}^{ref} as recommended by some previous studies, as well as, the default setting used in PLAXIS.

A numerical simulation using the Geotechnical Engineering Software (PLAXIS) is employed to analyze the soil response to the wall. First, the excavation is simulated by using PLAXIS 2D version 2020. Then, the vertical and horizontal boundary of the

Table.2 Ground profile of SCV

No	Soil layer	φ' (o)	c' (kPa)	E_{50}^{ref} (kPa)	m
0	Sand Fill	28.0	5.0	30,000	0.55
1a	Soft soil	24.7	6.7	6000	1.00
1b	Soft soil	24.7	6.7	7000	1.00
2a	Stiff clay	26.9	19.5	46,000	0.75
2b	Stiff clay	26.9	19.5	57,000	0.75
3a	Sand	26.0	5.2	64,000	0.55
3b	Sand	26.0	5.2	80,000	0.55
4	stiff clay	28.6	32.9	63,000	0.75

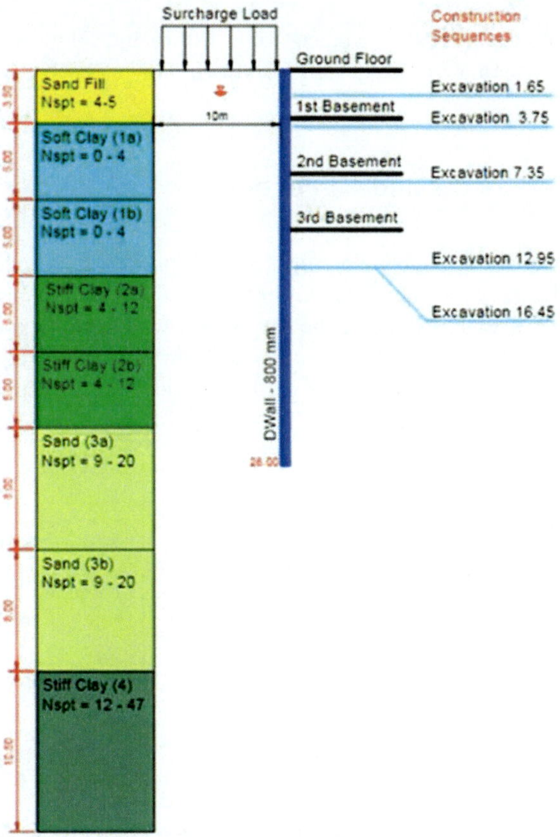

Fig. 4 Ground data and information of Sunrise City View project

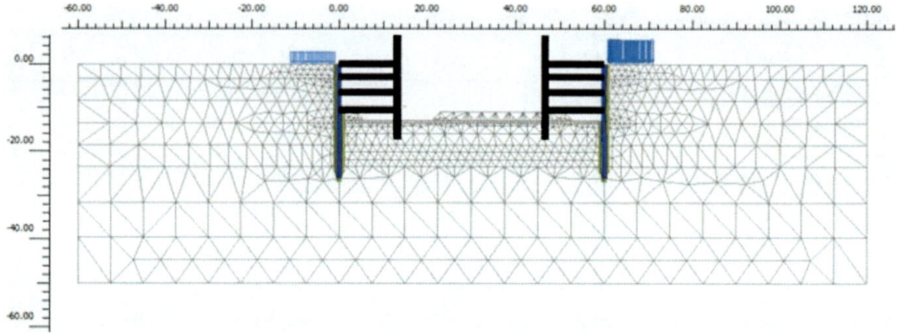

Fig. 5 Simulation SCV using PLAXIS 2D

calculated mesh is created, to be 180 m and 50 m, respectively. Figure 5 illustrates the calculated mesh of this problem.

3 Results and Discussion

The whole excavation consists of 10 phases, in which five stages excavation are in phases, which are the main impact on the lateral displacement diaphragm wall. The displacement value of the diaphragm wall at first excavation is quite large because soil layer 1 is thick and very soft, with N_{spt} almost equal to 0. The displacement of diaphragm wall increases gradually with the excavation process, namely excavation to 1.65 m, 3.75 m, 7.35 m, 12.95 m and 16.45 m (see Fig. 6). The maximum displace-

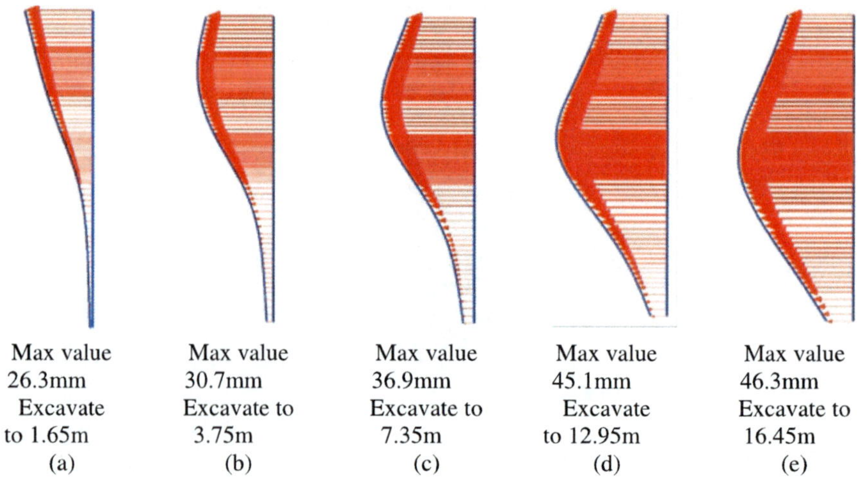

Fig. 6 Horizontal displacement diaphragm wall

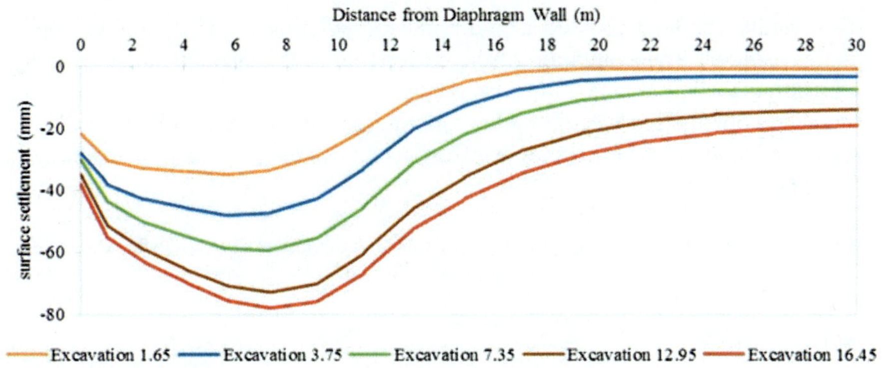

Fig. 7 Surface settlement

ment value is 46.3 mm at the final stage (foundation excavation). Simultaneously, there is a process of increasing the surface settlement. The largest settlement predicted at each stage is shown in Fig. 7. The ground settlement is estimated with the largest value, about 8 m from the diaphragm wall. The impact of excavation on the ground surface settlement decreases steadily with the distance from the diaphragm wall to the outside. This is illustrated in detail with a decrease in the surface settlement in Fig. 7.

It is noted that the largest value of the lateral displacement diaphragm wall of the final phase excavation is only 46.3 mm. This value of displacement guarantees that diaphragm wall can work against collapse under excavation. It means that this value is proper for a project using the semi top-down construction method.

4 Conclusion

Sunrise city view building is a special location, which combines many disadvantaged factors for excavation. This structure is constructed on basing thickly soft soil in an urban city and near a river. This leads to more meaningful simulations and predictions of what happens. Based on geotechnical engineering software (PLAXIS), an estimation method for displacement diaphragm wall is presented. With some modern features, a semi top-down approach is employed as an appropriate solution for the urban excavation procedure of this project. The soil and structural retaining wall response is described thoroughly by the advanced soil model (Hardening Soil model). This simulation can support many different conditions, namely soil characteristics, soil interaction, and other materials such as concrete or steel, water level, etc. Besides, it also provides the environment for simulation the real phenomena with analytical methodologies.

The results obtained provide a preliminary prediction of what might happen during excavation. The evaluation of the correlation of the characteristics (namely the diaphragm wall, the soil parameters and the excavation process) is the foundation for detailed construction calculations in the later stages of the problem. The semi top-down method shows efficiency because it reduces displacement of diaphragm wall during excavation.

Acknowledgements The authors acknowledge the financial support of VLIR-UOS TEAM Project, VN2018TEA479A103, 'Damage assessment tools for Structural Health Monitoring of Vietnamese infrastructures' funded by the Flemish Government.

vliruos SHARING MINDS, CHANGING LIVES With the support of THE BELGIAN DEVELOPMENT COOPERATION .be

References

1. Börgesson L (1996) ABAQUS In: Stephansson O, Jing L, Tsang C-F (eds) Developments in geotechnical engineering, vol 79. Elsevier, pp 565–570
2. Helwany S (2007) Applied soil mechanics with ABAQUS applications. Wiley, Berlin
3. Brinkgreve R, Kumarswamy S, Swolfs W, Waterman D, Chesaru A, Bonnier P (2016) PLAXIS 2016, PLAXIS bv, the Netherlands
4. P. G. E. Software (2020)
5. Tan Y, Huang R, Kang Z, Bin W (2016) Covered semi-top-down excavation of subway station surrounded by closely spaced buildings in downtown Shanghai: building response. J Perform Constr Facil 30(6):04016040. https://doi.org/10.1061/(ASCE)CF.1943-5509.0000892
6. Hongbing W (1998) Semi-top-down construction technology at the Jinanshi station of metro line No. 2 in Shanghai. Undergr Space
7. Tan Y, Zhu H, Peng F, Karlsrud K, Wei B (2017) Characterization of semi-top-down excavation for subway station in Shanghai soft ground. Tunnell Undergr Space Technol 68:244–261. 2017/09/01
8. Huang Z-H, Zhao X-S, Chen J-J, Wang J-H (2014) Numerical analysis and field monitoring on deformation of the semi-Top-Down excavation in Shanghai. In: New frontiers in geotechnical engineering, pp 198–207
9. Shen SL, Wang ZF, Cheng WC (2017) Estimation of lateral displacement induced by jet grouting in clayey soils. Géotechnique 67(7):621–630
10. Talha SB (2000) Deformation behaviour of a retaining wall for a deep basement excavation with semi-top down method, p 1049

Experimental Investigations and Numerical Simulations for the Seismic Assessment of a Masonry Building

Mariella Diaferio, Marilena Venerito, and Michele Vitti

Abstract In Europe, a significant part of the existing building stock is composed by masonry structures, and, in many cases, these structures are located in seismic areas. They were built based on the empirical expertise of their constructors, and, therefore, are characterised by several construction solutions, different materials, and a high vulnerability to earthquakes. Moreover, during their long service life these buildings have been subjected to damage, degradation, and interventions to adapt them to the new performance demands. All these aspects make quite difficult the evaluation of the safety level of masonry structure, and several approaches are still debated for the definition of accurate numerical models for seismic assessment purpose. The present paper discusses the performed investigations and the defined numerical model for the evaluation of the safety level of a masonry structure characterised by two floors, by the presence of wooden roofs and masonry vaults at the first level, and by a pitched roof whose elements are made of wood. The building is located in a seismic area and is utilised as a church. The analysis of the documentation, the geometrical survey and the visual and experimental investigations have been utilised to define its level of knowledge, and to calibrate a numerical model for vulnerability evaluation.

Keywords Masonry structure · Level of knowledge · Vulnerability · Numerical model

1 Introduction

Our cities are characterized by the existence of a wide masonry building stock which is still in use and that, particularly in some Countries like Italy, represents an historical and monumental patrimony. This heritage must be preserved for several reasons, as testimony of the architectural evolution but also for the effects of its out of service on

M. Diaferio (✉)
Polytechnic University of Bari, DICATECh, 70125 Bari, Italy
e-mail: mariella.diaferio@poliba.it

M. Venerito · M. Vitti
Structural Engineering, 70100 Bari, Italy

M. Abdel Wahab (ed.), *Proceedings of the 2nd International Conference on Structural Damage Modelling and Assessment*, Lecture Notes in Civil Engineering 204,
https://doi.org/10.1007/978-981-16-7216-3_6

tourism and on the society, as in many cases these structures are utilized for public functions.

However, this patrimony due to its long service life is characterized by the effects of material degradation due to the environmental action, damage patterns induced by dynamic loads and/or foundation settlements, etc.

The seismic events of the last decades [1–4] underlined the high vulnerability of the masonry patrimony, that represents the largest part of the south Europe city centers, and the necessity of preserving this patrimony firstly to avoid human loss, and also the failure of historical testimonies.

Thus, the reliable evaluation of the current safety level of a masonry structure has reached increasing attention by researchers for the social and economic implications, but also for the difficulties connected to the definition of a complete and widely applicable approach to the analysis of these structures [3, 5–7]. In fact, due to the high variability of construction typologies, of t he architectural style and mechanical characteristics, several approaches and applications are discussed in the technical literature.

The prerequisite for the evaluation of the current safety level of these structures against earthquakes and the design of possible effective interventions, is the acquisition of a good level of knowledge which is usually difficult to acquire due to the lack of documentation, the existence of changes of the original structure occurring over time, resulting from the necessity of adapting the structure to the new required performance, the occurrence of damage phenomena, etc.. In many cases these interventions are responsible of increasing the irregularity of the structure and their vulnerability to seismic actions.

Several techniques are available for the complete mechanical characterization of masonry structures, however the most reliable are too invasive and thus in many cases may not be adopted due to the necessity of preserving the structure, or are time-consuming and economically expensive.

Thus, in literature many researchers have proposed the use of the laser scanner for the acquisition of a reliable geometry [2, 8]. In [2] the geometrical survey by means of 3D Laser Scanner and the seismic vulnerability assessment of the baptistery of San Giovanni in Tumba located in Monte Sant'Angelo (Italy) is discussed. The paper shows the advantage of the Laser Scanner methodology, which is able to acquire a huge number of points within limited time intervals, allowing the definition of a complete and detailed 3D model. These characteristics justified the increase of the researchers' attention for the laser scanner with the aim of defining a refined 3D model of the cultural and historical heritage.

The present paper aims to highlight the relationship between a detailed survey, also including ad hoc chosen experimental tests, and the structural analysis of an irregular structure, by discussing in detail the procedure for the seismic vulnerability analysis of the S. Marco Evangelista Church in Bari (Italy). This structure has been chosen for its complexity and the presence of structural elements characterized by different behavior. It is discussed the acquisition of geometrical data by means of Laser Scanner, the seismic P-wave refraction method for soil characterization, the endoscopic tests to exactly evaluate the construction technology and geometrical features of masonry structural elements, the compressive tests on drilled cores and the results of the structural analysis on a Finite Element (FE) model both for static and dynamic loading conditions. The paper is organized as follows: in Sect. 2 the examined Church is described, Sect. 3 discusses in detail the experimental campaign carried out to increase the knowledge level of the structure. In Sect. 4 the definition of the finite element model of the Church is described and the results of the seismic analysis are illustrated in detail.

2 The San Marco Evangelista Church

The San Marco Evangelista Church is located in Bari (Apulia, Italy), and is an isolated masonry palace surrounded by masonry walls (Fig. 1). The building has plan dimensions of 16 m × 23 m, a maximum height of 10.5 m and it comprises two levels and an attic.

The San Marco Evangelista church was built during the VIII century and its original function was farmhouse. The visual inspection of the structure highlights that it was built in two periods, as two different materials can be recognized in the structure. In detail, the main nave of the church and the apse have been realized in the same period and with the same materials, while the remaining rooms are characterized by a more porous material, and it was reasonable built prior to the first cited part. This consideration is confirmed by the presence of an arch which is not coherent with the actual orientation of the floor, it can be hypothesized that in the original structure, which did not comprehend the actual nave and apse, the floors were oriented orthogonally to the actual direction, coherently with the distribution of the masonry walls, however, when the other parts of the structure were built, the floors were substituted and oriented differently (see Fig. 2). The Church presents cracks localized in the zones where the two materials are connected.

In the 1977 the town Council changed its function and established that the patronal rooms, the oil mill, and the stables were used for religious functions, then the bishop Ballestrero dedicated the structure to St Marco Evangelista.

Between the 1980 and the 1990, a retrofitting intervention was carried out on the structure consisting of the demolition of the floor located on the actual position of presbytery and subsequent reconstruction with a wooden floor.

Fig. 1 View of the San Marco Evangelista Church in Bari (Italy)

(a) (b)

Fig. 2 **a** Frontal view of the East façade; **b** Frontal view of South façade of San Marco Evangelista Church

During the 1985 due to a tornado, the roof of the building was significantly damaged and, thus, the church was subjected to an extensive intervention during which the roof was rebuilt with a wooden pitched roof. The historical research has not given any information on the interventions during the period between the building construction and the 1980.

Observed from the outside, the building appears composed by two rectangular plan parts, orthogonal to each other. The first rectangular part is oriented East–West while the second part is along the North–South direction and connects the liturgical hall directly with the sacristy.

The liturgical hall is at the ground floor in two adjoining rooms, which originally were used as oil mills, and that have a characteristic "L" shape with the altar placed at an angle in order to guarantee most of the visibility. Each of the two rooms is characterized by round arches that support two lowered barrel vaults orthogonal to each other, whose intersection determines a groin vault.

All the rooms at the ground floor are characterized by exposed stone walls except for the altar, while the rooms on the first floor are entirely plastered.

The parish office and some classrooms for pastoral activities are located at the first level.

An internal masonry staircase, consisting of two ramps supported by a barrel vault, and an external masonry staircase located on the main façade allow the access to the first floor, while the attic can be reached only by means of the internal staircase.

The Church presents barrel vaults and floors realized by planks and timber beams.

The attic is characterized by a trussed wooden roof supported by the walls and by central masonry pillars. To reduce the span lengths of the ridge beam, wooden struts are installed in the vertical plane.

The entrance to the liturgical hall is located on the East side of the building where two openings of different sizes are placed: the first with a lowered arch allows the entry to the liturgical hall, the second, which is smaller than the first one, allows entry to the sacristy and ancillary rooms. On the longest façade of the building there is a secondary entrance that allows lateral access to the liturgical hall.

3 Experimental Campaign

The lack of documentation regarding the geometric and structural characteristics of the Church, has imposed a detailed geometrical survey and experimental tests to improve the knowledge level and to perform the subsequent seismic vulnerability assessment.

In detail, a series of laser scanning surveys have been planned for defining a reliable finite element model of the structure.

This methodology was chosen as it allows a complete description of the structure, overcoming the main limit of the "traditional" survey whose accuracy is dependent from the choices of the operator, that unavoidably must discretize the structure, and in this way, inevitably effects the accuracy of the numerical model.

The use of 3D Laser Scanner provides a 3D point cloud data wh ich allows the definition of a high-resolution geometrical model of the investigated structure. This objective is obtained by merging many acquisitions obtained by changing the position of the 3D Laser Scanner. The three-dimensional point cloud data are later analyzed by inserting boundary and connectivity conditions which allow to define a refined geometrical model.

The adopted 3D Laser Scanner is Leica HDS 6100 which is a continuous wave phase-based TLS characterized by an acquisition velocity equal to 127,000 pts/s, it allows a large field of view (360° × 270°), ensuring at the same time a low beam divergence and a good measuring accuracy (a standard deviation on a single acquisition equal to 3 mm at 25 m and 5 mm at 50 m).

To acquire a complete and very detailed three-dimensional model of the San Marco Church a total number of 23 scans were collected, which have been correlated to each other by a series of reflective targets.

At the end of the process, the resulting aligned model of the San Marco Evangelista Church shown in Fig. 3 has been obtained, which has an average root mean square error of the overall alignment around 2 mm.

Fig. 3 Computer rendering of the 3D model of the San Marco Evangelista Church acquired by means of Laser Scanner

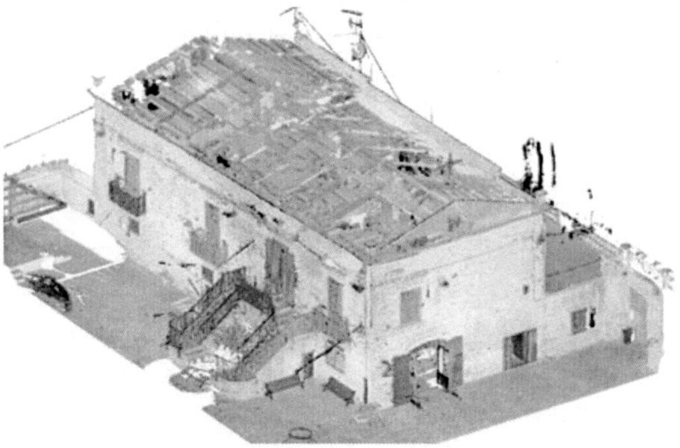

From the superimposition of the plans of the three floors, it is possible to deduce that the walls appear to have a continuity in elevation, the pillars of the attic, which support the wooden beams of the roof truss, are instead not perfectly aligned with under walls.

Moreover, endoscopic tests have been carried out to acquire detailed information regarding the stratigraphy and geometrical features of the walls and vaults. In detail, the Rigid micro Explorer was adopted which is a micro inspection camera with a diameter of 17 mm and a 320 × 240 RGB resolution, placed at the end of the endoscopic cable, equipped with an entirely digital platform that allows to record images and videos with a resolution of 640 × 480.

The endoscopic tests have been carried out on 7 masonry walls and a vault. The tests show that the masonry walls at the ground floor are realized by tuff bricks organized on two leaves whose dimensions vary along the perimeter of the building, while the walls inside the church are realized by a single leaf of tuff brick. The latest typology is also observed at the first floor, except for the central wall which is a double- leaf wall in tuff bricks.

To evaluate the mechanical parameters of the materials, two cylindrical specimens were extracted by means Diamond core drill IM250H, with an external diameter of 80 mm and internal diameter of 74 mm, and the two cores have been subjected to compression tests (Fig. 4). The evaluated compression strength was equal to 1.88 MPa.

Moreover, with the final aim of estimating the mechanical characteristics of the walls and in accordance with the Italian Code [9] and the 2019 Guidelines [10], the Masonry Quality Index (MQI) [11–13] method was applied which makes use of a visual survey of the faces and the cross section of wall panel and proceeds by verifying if it obeys the "rules of the art". The visual analysis makes use of 7 parameters (usually indicated as SM, SD, SS, WC, HJ, VJ, and MM). It is worth noting that the masonry quality influences differently the response to different actions, thus, the cited 7 parameters must be evaluated based on the acting loads. The seven parameters are:

1. MM Parameter—Quality of the Mortar/Contact between Masonry Units: the quality of mortar affects the transfer of stresses between masonry elements avoiding the concentration of stresses.

Fig. 4 Compression test on a drilled core

2. WC Parameter—Level of Connection between Adjacent Wall Leaves/Headers: the global behavior of the wall, particularly the compressive and out-of-plane ones, are highly influences by the headers connecting the wall leaves, because they allow the distribution of the gravitational loads on the whole wall transversal section.
3. SS Parameter—Shape of the Masonry Units: the use of perfectly cut elements affects the response to horizontal in-plane loads.
4. SD Parameter—Dimensions of the Masonry Units: this parameter is based on the observation that the static and dynamic behavior of the wall improves as the ratio between the brick length and the wall width increases.
5. VJ Parameter—Staggering of Vertical Mortar Joints: this parameter takes into account the effect on the wall static and dynamic behavior of stagger properties of vertical joints.
6. HJ Parameter—Horizontality of Mortar Bed Joints: the presence of continuous and horizontal layers of mortar allows to better distribute the gravitational loads on the panel transversal section and improves the response under seismic actions.
7. SM Parameter—Mechanical Characteristics and Quality of Masonry Units: this parameter considers the brick mechanical properties and conservation state.

On such basis, the MQI must be evaluated for tree different load conditions: vertical loading, horizontal in-plane loading and horizontal out-of-plane loading, the expression is the following:

$$MQI = m \times SM \times (SD + SS + WC + HJ + VJ + MM) \tag{1}$$

where m is a factor introduced to consider the influence of the quality of the mortar, which is assumed equal to 1 except for very-weak mortar, when it is set equal to 0.7.

Based on the MQI values, three different masonry categories are individualized, denoted by A, B, C whose behavior worsens from A to C.

The masonry walls of the San Marco Church have been classified as s hown in Table 1, where NF stands for Not Fulfilled, PF for Partially Fulfilled, and F for Fulfilled (see Fig. 5), while the mechanical properties have been evaluated in accordance with the prescription of the Italian Codes for the squared softstone masonry.

To characterize the soil, on-site tests have been performed; in detail, the seismic P-wave refraction method has been adopted which evaluates the compression waves generated b y a seismic source located at a known distance from the investigated site, this test was chosen as these waves are more clearly identifiable. In the seismic refraction survey, the transducers detect the refracted waves traveling at higher speeds. Knowing the first arrival times and the geophone-source distance, and by analyzing the curves of the first arrivals to each transducer (travel time), it is possible to determine the speed of the various layers and to produce a seismic stratigraphy to be correlated to the geological formations or discontinuities present in the investigated soil.

Table 1 Evaluation of MQI for a masonry panel of the San Marco Evangelista Church

		Vertical loading (V)	Horizontal in-plane Loading (I)	Horizontal out-plane Loading (O)
MM	NF	0	0	0
WC	PF	1	1	1.5
SS	F	3	2	2
SD	F	1	1	1
VJ	PF	0.5	1	0.5
HJ	F	2	1	2
SM	PF	0.7	0.7	0.7
	QMI	5.25	4.2	4.9
Masonry category		A	B	B

Fig. 5 Detail of a masonry panel for the evaluation of MQI

The method is based on the observation that the P-wave speed through a medium depends on the physical properties (i.e. rigidity, density, saturation) and degree of homogeneity of the material.

In the here examined case, the seismic P-wave refraction tests have been performed adopting a length of the survey line equal to 50 m, located parallel to the east façade of the Church.

The source of P waves was an instrumented impact hammer with a weight equal to 8 kg which hitts vertically on an aluminum plate suitably placed on the ground, for the prevailing generation of rich high frequency P waves, with repeatable and directional waveforms.

The waves were acquired by means of 24 electromagnetic geophones with moving-coil characterized by a natural frequency equal to 4.5 Hz. The geophones were installed at a constant distance of 2 m and the frequency acquisition was set equal to 256 Hz. To obtain a complete description of the soil, 8 different shot points were chosen: two located at the ends of the acquired line and the other 4 shot points in correspondence of the 5th, 10th, 15th and 20th geophone, respectively.

The data processing was carried out by means of software RAYFRACT [14].

In Fig. 6 the acquired seismogram for the case in which the shot point is at the start point of the survey line is shown, while in Table 2 the first arrival times of P-waves at the surface of the earth for each geophone are presented.

The results of the soil inspection are shown in Fig. 7 where a contour plot shows the acquired wave velocity profile at varying depth from the ground level (y axis) and along the survey direction (x axis). The plot shows that the subsoil consists of 2 basic seismic stratigraphic units from the ground level:

- The first seismic stratigraphic unit UsC (see Fig. 7) extends up to an average depth between 1.3 ÷ 2.8 m from the ground level with an almost stratiform structure. This seismic layer is characterized by V_p values ranging between 500 ÷ 900 m/s, with minimum values recorded at a distance along the survey direction greater that 22 m, while the minimum values are reached beyond the cited distance. The boundary, which divided this unit from the underlying seismic layer, appears clearly marked as it is characterized by very high gradient values.

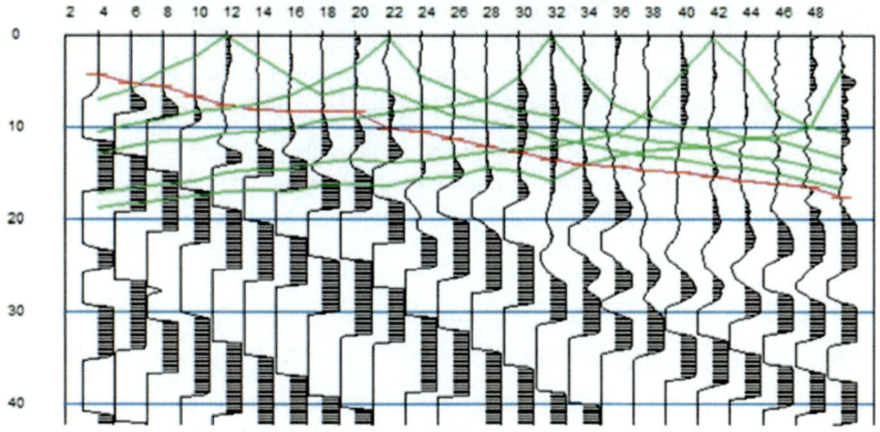

Fig. 6 Seismogram for the shot point at the beginning of the survey line

Table 2 Seismic P-wave refraction tests: first arrival times for each geophone with the shot point at the start point of the survey line

Geophone	1	2	3	4	5	6	7	8	9	10	11	12
[ms]	4.21	5.15	5.62	6.61	7.55	8.02	8.31	8.31	8.31	10.11	10.53	11.16
Geophone	13	14	15	16	17	18	19	20	21	22	23	24
[ms]	12	12.6	13.5	14.1	14.2	14.8	15	15.4	15.9	16.15	16.47	17.52

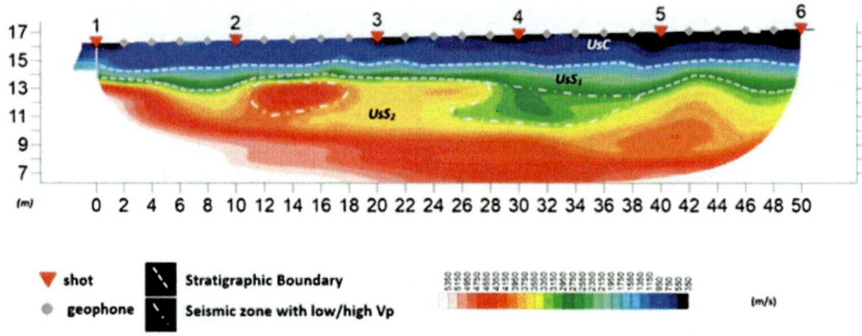

Fig. 7 Seismic stratigraphy: x axis distance [m] along the survey direction and y axis distance along the vertical direction [m]

- The second seismic stratigraphic unit UsS (see Fig. 7) shows a clear increase in the detected speeds and an uneven distribution of the values. This unit can be further divided into 2 subunits: the first unit UsS_1, characterized by a variable thickness ranging between $1.0 \div 2.0$ m, reached the maximum values in the range of distance along the survey direction equal to 28–40 m. This zone manifests V_p values ranging between 1450 m/s and 2400 m/s with values increasing with depth. The second seismic stratigraphic unit UsS_2 is characterized by V_p values greater than 3300 m/s with an uneven distribution, moreover, at a depth of 7 m from the ground level, the V_p values increase ranging between 4500 m/s and 5700 m/s.

In accordance with the experimental results and with the Italian Codes prescriptions [9, 10], the category A (Rock or other rock-like geological formation, including at most 3 m of weaker material at the surface) was assumed for the soil, while the topographic category was assumed equal to T1.

4 Numerical Model

To evaluate the safety level of the San Marco Evangelista Church, a reliable finite element model was defined based on the results of the experimental campaign and utilized to estimate the response to seismic action compatible with the seismic hazard of the zone. To this aim, a response spectrum analysis was carried out to assess the seismic vulnerability.

The numerical model assumes that the walls have a homogeneous stratigraphy whose thickness is imposed based on the endoscopic results. The vertical walls and the vaults have been modelled by means of 7338 shell elements, while the columns at the attic and the roof elements have been modelled through 861 beams, and the floors have been modelled by means of 502 elements which distribute the loads. The whole model is characterized by 7664 points; in Fig. 8 the 3D view of the model

(a) (b)

Fig. 8 3D view of the finite element model of the San Marco Evangelista Church: **a** south view; **b** north view

is shown, the different colors are related to the adoption of different geometrical thicknesses.

The experimental compressive strength of tuff is equal to 1.88 MPa, and the experimental compressive strength of mortar is equal to 2.50 MPa, thus the characteristic compressive strength of masonry is obtained as follows:

$$f_k = K f_b^{0.7} f_m^{0.3} \tag{2}$$

where K = 0.45; thus, f_k=0.92 MPa.

The knowledge level is equal to LC3 with a confidence factor equal to 1 [9, 10], moreover, in accordance with such codes, the mechanical properties have been increased by a coefficient equal to 1.2, and the final mechanical parameters are shown in Table 3.

The analysis has been carried out by adopting a snow load equal to 0.8 KN/m^2, wind load 0.65 KN/m^2, moreover, for the wooden floors the structure's self-weights have been set equal to 0.08 KN/m^2 and non-structural permanent loads equal to1.00 KN/m^2. The live load at the ground and first level has been set equal to 2.00 KN/m^2, and at the attic equal to 0.5 KN/m^2. The evaluation of self-weights of vault abutments has been carried out by considering the variation of the height along the vault, thus,

Table 3 Mechanical parameters adopted for the FE model

Material	Young's modulus [MPa]	Modulus $E_{0.05}$[MPa]	Shear modulus [MPa]	Density [kg/m^3]
Regular softstone masonry	2112	–	845	1580
Wood C18	10,500	5999	660	375
Glued laminated timber	11,500	9599	650	385

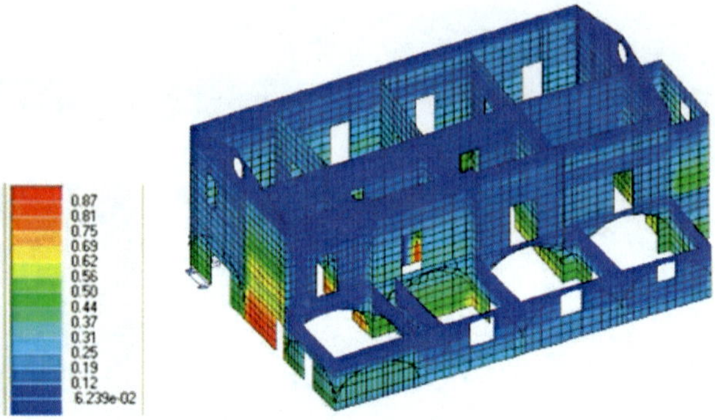

Fig. 9 Results of the safety checks related to the axial forces and out-of-plane bending moment induced by static loading condition

different values of load have been adopted depending on the position of the considered shell element on the vault.

In the first step of the analysis, a static analysis has been carried out to estimate the safety level of the Church in the usual loading conditions. In Fig. 9 the plot of safety checks related to the axial forces and out-of-plane bending moment is shown.

In accordance with the Italian Building Codes [9, 10], due to the irregularities of the structure both in elevation a nd in plan, a behavior factor $q = 2.25$ for the horizontal design spectra and $q = 1.5$ for the vertical direction have been adopted.

The response spectrum analysis has been carried out to evaluate the seismic vulnerability of the church. In Fig. 10 the plot of the safety checks related to the axial forces and in-plane bending moment due to the worst load conditions is shown, in this case the load combination considers the seismic action too.

For the same load combination, in Fig. 11 is plotted the results of the safety checks related to the shear action. It is worth noting that the structure is subjected to shear forces not compatible with its resisting capacities.

Fig. 10 Results of the safety checks related to the axial forces and in-plane bending moment induced by the worst seismic load condition

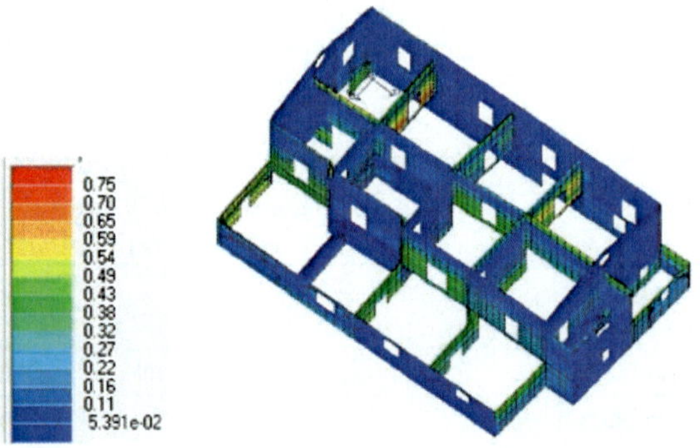

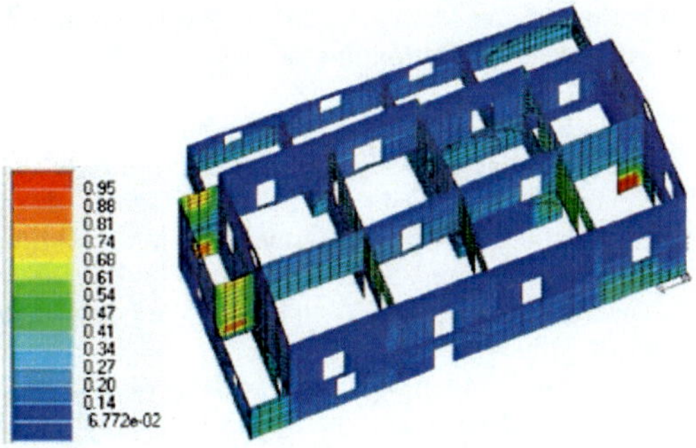

Fig. 11 Results of the safety checks related to the shear forces induced by the worst seismic load condition

5 Conclusion

The paper presents the survey and the seismic vulnerability evaluation of San Marco Evangelista Church in Bari (Italy), which dates back to the VIII century, having cultural value and that can be considered representative of the Italian historical patrimony.

Due to the irregularities both in plan and in elevation and the lack of information regarding the original configuration of the structure, an extensive campaign was carried out. The present paper has shown the benefits of the use of Laser Scanner methodology for defining the exact geometric configuration, and the relationship with the definition of an accurate finite element model. The experimental campaign allowed the detailed characterization of the masonry structural elements and of the soil.

The experimental analysis and the definition of an accurate numerical model highlighted that the San Marco Evangelista Church capacity may be lower than the demand when it is subjected to seismic actions, and that has an average vulnerability degree.

Acknowledgements The project Politecnico di Bari FRA 2016 "La tutela del patrimonio storico-artistico: rappresentazione, analisi di vulnerabilità, risanamento e valorizzazione" is acknowledged for the support given to the present research

References

1. Blyth A, Di Napoli B, Parisse F, Namourah Z, Anglade E, Giatreli AM, Rodrigues H, Ferreira T (2020) Assessment and mitigation of seismic risk at the urban scale: an application to the historic city center of Leiria, Portugal. Bulletin Earthq Eng 18:2607–2634
2. Fortunato G, Funari MF, Lonetti P (2017) Survey and seismic vulnerability assessment of the Baptistery of San Giovanni in Tumba (Italy). J Cult Herit 26:64–78
3. Valente M, Milani G (2016) Nonlinear dynamic and static analyses on eight historical masonry towers in the northeast of Italy. Eng Struct 114:241–270

4. Formisano A, Marzo A (2017) Simplified and refined methods for seismic vulnerability assessment and retrofitting of an Italian cultural heritage masonry building. Comput Struct 180:13–26
5. Diaferio M, Foti D, Sabbà MF, Lerna M (2021) A procedure for the seismic risk assessment of the cultural heritage. Bulletin Earthq. Eng 19(2):1027–1050
6. Lourenco PB, Oliveira DV, Leite JC, Ingham JM, Modena C, da Porto F (2013) Simplified indexes for the seismic assessment of masonry buildings: international database and validation. Eng Fail Anal 34:585–605
7. Diaferio M, Venerito M, Vitti M (2018) Experimental testing and numerical analysis of a barrel vault. Structures 15:138–151
8. Guarnieri A, Milan N, Vettore A (2013) Monitoring of complex structure for structural control using terrestrial laser scanning (TLS) and photogrammetry. Int J Archit Herit 7:54–67
9. Ministerial Decree of Public Works (M. D.) (17/01/2018). New technical codes for constructions. Official Gazette of the Italian Republic n. 42 published on 2018, February 20th.
10. Ministerial Circular 21/01/2019. Instructions for the application of the "New Technical Code for Constructions". Official Gazette of the Italian Republic published on 2019.
11. Borri A, De Maria A (2009) Scheda di Valutazione dell'IQM (Indice di Qualità Muraria); Rete dei Laboratori Universitari di Ingegneria Sismica (RELUIS): Naples, Italy
12. Borri A, Castori G, Corradi M, De Maria A (2015) A method for the analysis and classification of historic masonry. Bull Earthq Eng 13:2647–2665
13. Borri A, Corradi M, Sisti R, De Maria A (2018) Calibration of a visual method for the analysis of the mechanical properties of historic masonry. Int J Struct Integrity 11:418–427
14. Rayfract® Seismic Refraction/Borehole Tomography software, Canada https://rayfract.com

A Portable Scanning Device for Local Vibration Testing of Concrete Structures

Tatsuro Murakawa, Hideki Naito, Yusuke Fujisaku, Kohko Inaba, and Takatada Takahashi

Abstract In this study, the effectiveness of local vibration testing using a portable scanner and machine learning for non-destructive inspection of concrete structures was examined. Using a portable vibrator and laser vibrometer, local through-thickness vibration tests were conducted on specimens of concrete beams and a railway track containing voids. Comparisons between frequency response functions measured over intact and void regions within structural concrete showed wave damping around the void. Moreover, a large amount of frequency response functions was obtained with the laser vibrometer in the scanning testing. With the measured data, Support Vector Machine (Kernel method) was used for detecting voids within the beams and beneath the railway track slab. Through its analysis of frequency response functions, the measured data of the intact or void condition was classified with a percent accuracy of 70%. This result indicates promising usage of the proposed method: utilizing a portable vibrator, laser vibrometer, and machine learning for non-destructive inspection of concrete structures.

Keywords RC Structures · Non-Destructive Testing · Machine Learning · Damage Detection

1 Introduction

Maintenance of civil structures is essential around the world. Structural health monitoring of bridges during their service periods has been established [1]. Non-destructive inspection vehicles with sensors and testing equipment have been developed [2], as high accurate sensors and large storage data systems can be utilized at lower costs. In addition, the use of AIs and automation technology for the inspection of civil structures [3, 4] is expected to show improvement growth in the near future.

T. Murakawa · H. Naito (✉) · Y. Fujisaku
Department of Civil and Environmental Engineering, Tohoku University, Sendai, Japan
e-mail: hideki.naito.c2@tohoku.ac.jp

K. Inaba · T. Takahashi
Track Technology Division, Railway Technical Research Institute, Tokyo, Japan

M. Abdel Wahab (ed.), *Proceedings of the 2nd International Conference on Structural Damage Modelling and Assessment*, Lecture Notes in Civil Engineering 204,
https://doi.org/10.1007/978-981-16-7216-3_7

Damages (e.g., internal cracks, voids, delamination between asphalt pavement and concrete slab) are caused within concrete structures under severe loads and environmental impacts. Non-destructive inspection methods are needed to detect the hidden damage within structures, since the damage greatly decreases their structural performance. However, vibration and displacement monitoring of bridges has shown challenges in detecting this type of damage, since global vibration characteristics (e.g., resonant frequency, damping ratio) change slightly with local damage [5, 6]. In addition, most conventional testing methods (e.g., ultrasonic testing [7, 8], impact echo measurement [9], thermographic images [10], and electro-magnetic radar [11, 12]) detect damage near-surface of concrete structures, but face difficulty in exposing internal cracks, voids, and delamination located far beneath surfaces.

On the other hand, local through-thickness vibration testing using an electric–magnetic vibrator has been useful in identifying internal damage within concrete structures [13, 14]. In this method, a local vibrator is excited, as shown in Fig. 1. Damages (e.g., internal cracks, voids, and delamination) around the excitation site are reflected in measured changes in frequency responses, which includes reflection and diffraction of waves around the void. However, local vibration testing is less efficient when it comes to inspecting large structures. Its fixed testing style requires numerous measurement locations and a lengthy test duration. Therefore, it would be beneficial to develop a method that improves the overall efficiency of civil structure inspection.

In this study, a portable scanning device for the inspection of concrete structures was investigated. A vibrator and laser vibrometer were gathered in a movable wagon, as shown in Fig. 1. White noise was applied through a wheel of the wagon from the top surface of the structures, where then a response waveform around the excitation location was measured with a laser velocity meter. In the scanning testing, a large amount of frequency response functions could be obtained from the waveforms through a fast Fourier transform (FFT). The large dataset of frequency responses gathered is useful for machine learning in classifying measurement locations to an intact or damaged condition.

Local vibration tests were conducted on concrete beam specimens containing a void. Frequency response functions obtained from fixed and scanning testing were

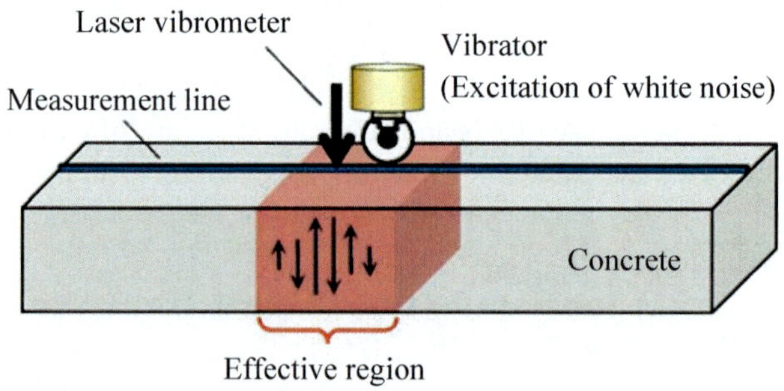

Fig. 1 Schematic diagram of local (through-thickness) vibration testing

compared. Damage detection through machine learning was verified. This method was then applied in detecting a void beneath of a railway track slab.

2 Concrete Beam Specimens

2.1 Experimental Program

Dimensions of concrete beam specimens containing a void are shown in Figs. 2 and 3. Foam sheets with a thickness of 5 mm were placed before concrete casting. The sizes of the sheets were 200 mm in length and 150 mm in width. Steel bars were not arranged in the beams. The material properties of concrete were obtained from test cylinders (with a diameter of 100 mm and height of 200 mm), the speed of sound was 4000 m/s, and the mass density was 2350 kg/m^3.

The sides of the beams were supported by wooden sleepers. Local vibration tests were conducted on the beams from the top side, using a wagon with a vibrator and

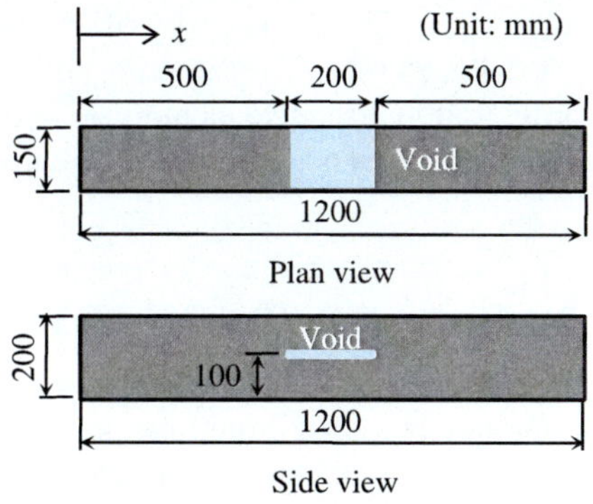

Fig. 2 Dimension of Specimen C

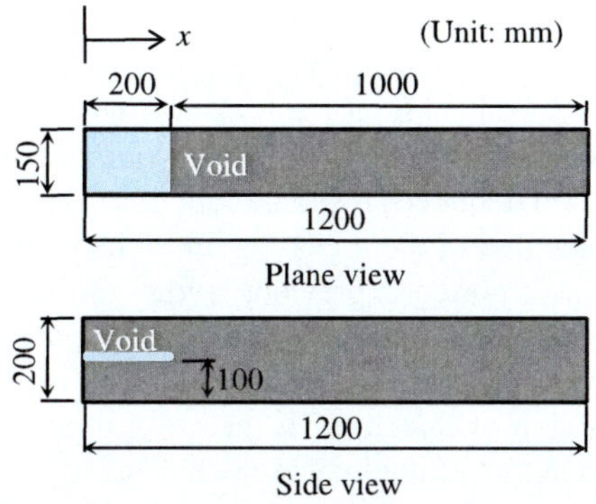

Fig. 3 Dimensions of Specimen E

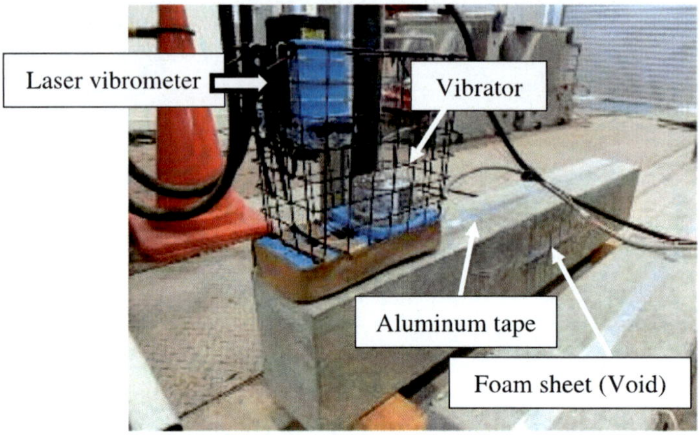

Fig. 4 Local vibration testing on Specimen C

laser vibrometer, as shown in Fig. 4. Aluminum tape with a width of 20 mm was pasted on the central line of the beams to better visualize the reflected laser light.

The vibrator was set to cover a frequency range of 1000 to 20,000 Hz. The acceleration power spectrum density (PSD) was 0.5 $(m/s^2)^2/Hz$, and the root mean square (RMS) was 98 m/s^2. The sampling period of the laser vibrometer was 20 μs. In other words, 50,000 discrete data plots of velocities were recorded per second. This large collection of data is beneficial to machine learning, as described later.

2.2 *Damage Detection in Fixed Vibration Tests*

Frequency response functions. In fixed vibration testing, measurement sites were set at the center of intact and void regions, as shown in Figs. 2 and 3. Response waveforms were measured during the 10 s application of white noise. Frequency response functions between 1000 and 20,000 Hz were calculated from 512 data plots taken from the waveforms, where the discrete step of frequency was 97 Hz.

Averaged frequency response functions were compared between the intact region of Specimen E (at $x = 600$ mm) and the void region of Specimen C (at $x = 600$ mm), as shown in Fig. 5. To reduce variation in amplitudes, the plots were averaged from 1000 frequency response functions over 10 s. Amplitude at the void region was higher than that of the intact region in the fixed testing, as shown in Fig. 5. This trend was also reported in previous studies on RC members [13, 14].

Averaged frequency responses at the side of the beams are shown in Fig. 6. They were measured at the intact region of Specimen C (at $x = 200$ mm) and the void region of Specimen E (at $x = 150$ mm). Amplitude at the void region was higher than that of the intact region. This result was similar to that shown in Fig. 5.

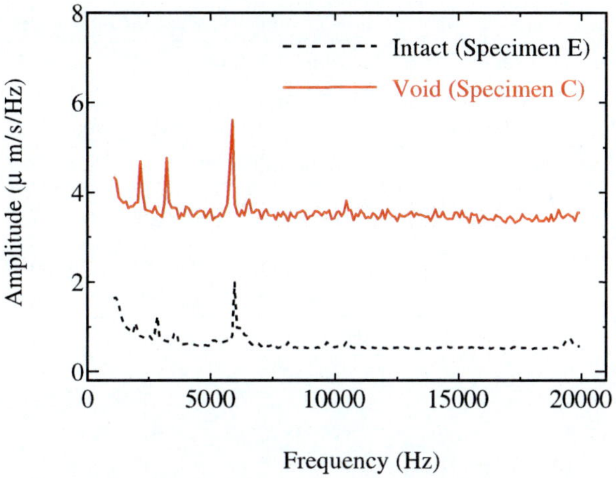

Fig. 5 Frequency response functions at the center of the beams (fixed testing)

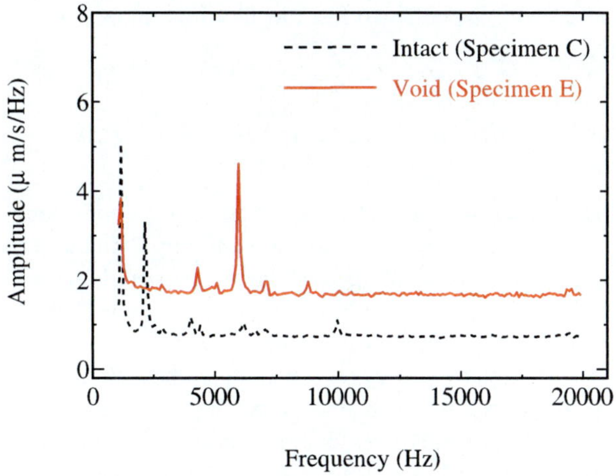

Fig. 6 Frequency response functions at the side of the beams (fixed testing)

Machine Learning. Frequency response functions were used as features in machine learning to classify measurement locations to the intact or void condition. Scikit-learn is a well-known open-source code on machine learning [15]. In this study, Support Vector Machine (Kernel method) was used as a classifier. In dataset preprocessing, features were given from the average of 10 frequency response functions between 1000 and 20,000 Hz, where each function was calculated by a FFT from 512 discrete data plots from waveforms measured with the laser vibrometer.

The number of data points used for machine learning is shown in Table 1. 80 and 20% of the dataset were used for learning and blind validation data, respectively. From

Table 1 Frequency responses used for machine learning (fixed testing on the beams)

Classes	Measurement sites in Specimen C (mm)	Measurement sites in Specimen E (mm)	Number of frequency response functions
Intact	$x = 200, 400, 800, 1000$	$x = 400, 600, 800, 1000$	800
Void	$x = 550, 600, 650$	$x = 0, 50, 100, 150$	700

Table 2 Confusion matrix and accuracy from machine learning (fixed testing on the beams)

Classes	Number of intact predictions	Number of void predictions	Percent accuracy (%)
Intact	206	32	87
Void	28	184	

the verification results, a confusion matrix and prediction accuracy from machine learning are shown in Table 2. Machine learning detected measurement locations to the intact or void condition with a percent accuracy of 87%. This result indicates that machine learning is effective in evaluating internal damage using a dataset of frequency response functions obtained from local vibration testing.

2.3 *Damage Detection in Scanning Vibration Tests*

Frequency response functions. Local vibration tests were conducted with the scanning wagon along the measurement line, as shown in Fig. 4. In the scanning testing, the wagon had a speed of 80 or 50 mm/s over the intact or void regions, respectively. The settings of the vibrator and laser vibrometer were the same as the fixed vibration testing described in Sect. 2.2. Each measurement case had three test trials.

Figure 7 shows averaged frequency response functions at the center of the beams. In each test trial, responses were obtained over the intact region of Specimen E (between $x = 680$–920 mm) and the void region of Specimen C (between $x = 525$–675 mm) for three seconds. In comparing Figs. 7 and 5, it is shown that the resulting frequency responses between the scanning testing and those of fixed testing were quite different. The amplitude in the scanning testing was around 10 times of that of fixed testing. Moreover, the amplitude decreased as frequency increased. It is thought that vibration was introduced by the moving wagon wheels. However, observations appeared different from those of white noise.

Figure 8 shows averaged frequency response functions at the side of the beams. They were obtained over the intact region of Specimen C (between $x = 80$–320 mm) and the void region of Specimen E (between $x = 25$–175 mm) for three seconds. The relationships between frequency and amplitude shown in Figs. 7 and 8 were similar. This indicates the possibility of wave damping around voids within concrete. However, further studies are necessary to interpret these results.

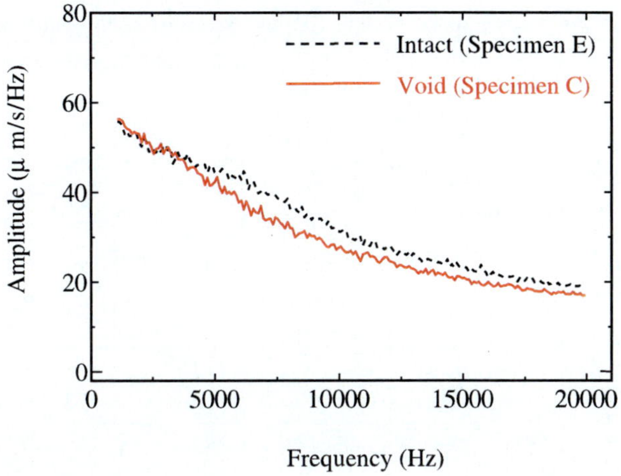

Fig. 7 Frequency response functions at the center of the beams (scanning testing)

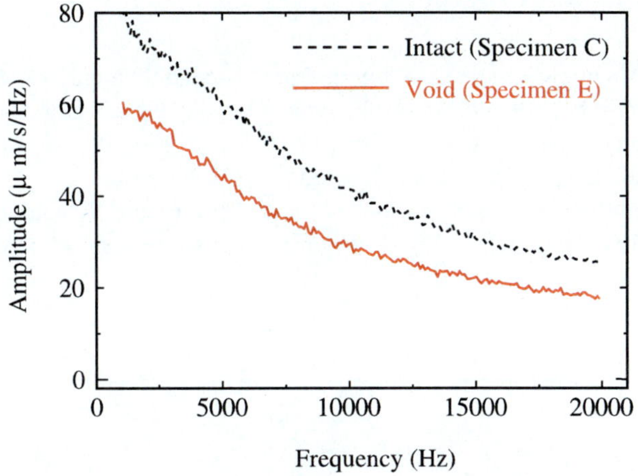

Fig. 8 Frequency response functions at the side of the beams (scanning testing)

Machine Learning. Damage detection from the scanning testing was examined using machine learning (Kernel method). Preprocessing of a dataset was the same as that of the fixed testing described in Sect. 2.2. The number of data points used for machine learning is shown in Tables 3. Additionally, 80 and 20% of the dataset were used for training and blind verification data, respectively.

A confusion matrix and prediction accuracy are shown in Table 4. From the scanning testing, machine learning classified measurement sites to the intact or void conditions with a percent accuracy of 73%. This accuracy is lower than that of fixed testing, as given in comparisons of Table 2 and Table 4. This indicates a decrease in evaluation accuracy with scanning inspection.

Table 3 Frequency responses used for machine learning (scanning testing on the beams)

Classes	Measurement sites in Specimen C (mm)	Measurement sites in Specimen E (mm)	Number of frequency response functions
Intact	$x = 80$ to 320	$x = 680$ to 920	180
Void	$x = 525$ to 675	x = 25 to 175	180

Note: Each measurement line had three test trials

Table 4 Confusion matrix and accuracy from machine learning (scanning testing on the beams)

Classes	Number of intact predictions	Number of void predictions	Percent accuracy (%)
Intact	39	15	73
Void	14	40	

3 Railway Track Slab

3.1 Experimental Program

A railway track sample was prepared with a precast slab. Dimensions and sectional view are shown in Figs. 9 and 10, respectively. Cement asphalt mortar (CA mortar) was filled between the RC slab and bed, as shown in Fig. 10. An unfilled (void) region with a thickness of 5 mm was formed beneath the slab.

Measurement lines were set on Lines H (intact region) and V (void region) in the running direction, as shown in Fig. 9. Figure 11 shows scanning vibration testing

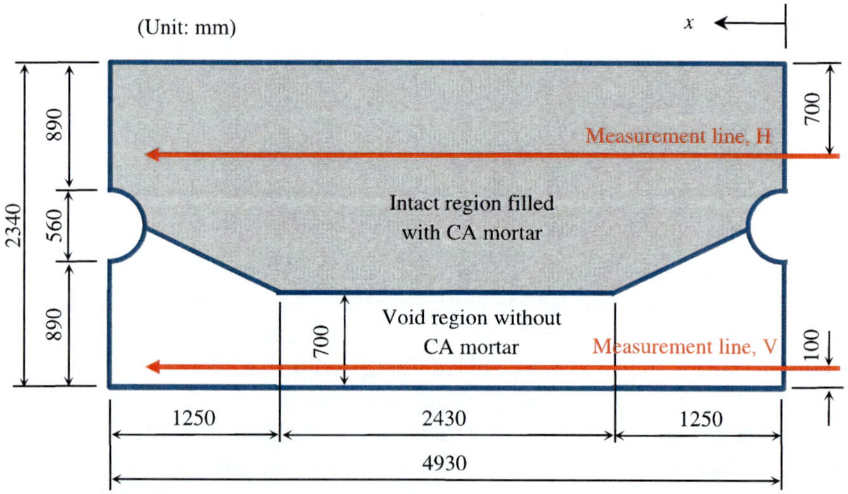

Fig. 9 Dimensions of the railway track slab

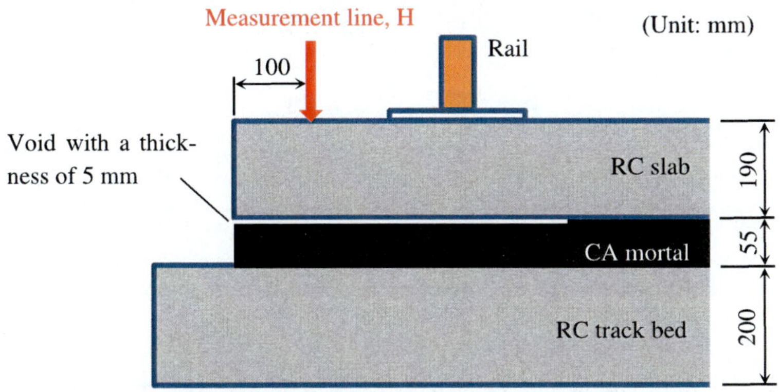

Fig. 10 Sectional view of the railway track slab

Fig. 11 Local vibration testing on the railway track slab

on the slab with the inspection wagon. The speeds of the wagon were 120 and 160 mm/s on Lines H and V, respectively. Basically, the settings of the vibrator and laser vibrometer were the same as the experiments on the beams described in Sect. 2. Aluminum tape was pasted on the measurement lines to better visualize the laser light reflection. Each measurement line had three test trials.

3.2 *Frequency Response Functions and Machine Learning*

Averaged frequency response functions obtained on Lines H and V were compared in Fig. 12, which gives an average of measured frequency responses over 25 s.

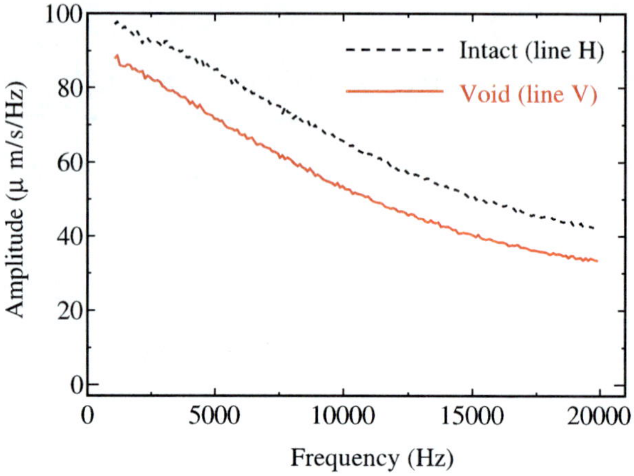

Fig. 12 Frequency response functions of the slab (scanning testing)

The frequency responses are similar to those of the scanning testing on the beams described in Sect. 2.3. Wave damping (decrease of amplitude) over the void region (Line V) is shown in Fig. 12.

Efficacy of machine learning for detecting a void beneath the slab was examined. Preprocessing of a dataset was the same as that described in Sect. 2. The number of data points used for machine learning is shown in Table 5. Further, 80% and 20% of the dataset were used for learning and blind verification data, respectively.

A confusion matrix and prediction accuracy are shown in Table 6. The machine learning method classified measurement sites to the intact or void region with a percent accuracy of 70%. This predicted accuracy coincides with the result for the beams shown in Table 4. From these observations, there indicates a possibility of utilizing this test method for detecting voids beneath railway track slabs.

Table 5 Frequency responses used for machine learning (scanning testing on the slab)

Classes	Measurement sites on Line H (mm)	Measurement sites on Line V (mm)	Number of frequency response functions
Intact	$x = 360$ to 4570	–	750
Void	–	$x = 890$ to 4040	750

Note: Each measurement line had three test trials

Table 6 Confusion matrix and accuracy from machine learning (scanning testing on the slab)

Classes	Number of intact predictions	Number of void predictions	Percent accuracy (%)
Intact	147	79	70
Void	56	167	

4 Conclusions

In this study, effectiveness of scanning vibration testing and machine learning for detecting damage of concrete structures was examined. From the experimental trials, the below conclusions and findings were obtained;

(1) A portable scanning device, a vibrator and laser vibrometer, can excite a local vibration in concrete beams and railway track slab. A large collection of frequency response functions along measurement lines can be obtained from the test.
(2) In the scanning vibration testing, wave propagation was damped around a void region of the beams and slab.
(3) Scanning vibration testing and machine learning are useful in detecting voids within concrete beams and beneath railway track slabs. Machine learning produced predictions for classifying intact or void conditions with a percent accuracy of 70%.

Though experimental, these conclusions and findings indicate the effectiveness of scanning vibration testing and machine learning for inspection of concrete structures. Additional investigations on vibration characteristics resulting from excitation by a running wagon, as well as damping of waves around a void, are needed.

References

1. Wong KY (2004) Instrumentation and health monitoring of cable-supported bridges. Struct Control Health Monit 11(2):91–124
2. Hugenschmidt J (2002) Concrete bridge inspection with a mobile GPR system. Constr Build Mater 16(3):147–154
3. Prasanna P, Dana KJ, Gucunski N, Basily BB, La HM, Lim RS, Parvardeh H (2016) Automated crack detection on concrete bridges. IEEE Trans Autom Sci Eng 13(2):591–599
4. Kim H, Ahn E, Shin M, Sim SH (2019) Crack and noncrack classification from concrete surface images using machine learning. Struct Health Monit 18(3):725–738
5. Salawu OS (1997) Detection of structural damage through changes in frequency: a review. Eng Struct 19(9):718–723
6. Carden EP, Fanning P (2004) Vibration based condition monitoring: a review. Struct Health Monit 3(4):355–377
7. Choi P, Kim DH, Lee BH, Won MC (2016) Application of ultrasonic shear-wave tomography to identify horizontal crack or delamination in concrete pavement and bridge. Constr Build Mater 121:81–91
8. Azari H, Nazarian S, Yuan D (2014) Assessing sensitivity of impact echo and ultrasonic surface waves methods for nondestructive evaluation of concrete structures. Constr Build Mater 71:384–391
9. Zhu J, Popovics JS (2007) Imaging concrete structures using air-coupled impact-echo. J Eng Mech 133(6):628–640
10. Gu JC, Unjoh S, Naito H (2020) Detectability of delamination regions using infrared thermography in concrete members strengthened by CFRP jacketing. Compos Struct 245:11 published online 112328
11. Bungey JH (2004) Sub-surface radar testing of concrete: a review. Constr Build Mater 18(1):1–8

12. Yu T, Twumasi JO, Le V, Tang Q, D'Amico N (2017) Surface and subsurface remote sensing of concrete structures using synthetic aperture radar imaging. J Struc Eng 143(10):11, published online 04017143
13. Naito H, Bolander JE (2019) Damage detection method for RC members using local vibration testing. Eng Struct 178:361–374
14. Naito H, Sugiyama R, Bolander JE (2019) Local vibration testing and damage evaluation for RC bridge decks. J Struct Eng 146(9):9, published online 04020168
15. Scikit-learn Homepage, https://scikit-learn.org, last accessed 2021/3/13

Detecting Delamination of Pavement Layers in Airfield Runways Using Local Vibration Testing and Machine Learning

Yusuke Fujisaku, Hideki Naito, Yu Shirai, Takuya Maeshima, Sonoko Ichimaru, and John E. Bolander

Abstract Delamination between the layers of asphalt pavement in airfield runways is studied by local vibration testing and machine learning. Debonding between pavement layers is modeled by placing sand or paper between the layers in the field experiment. The vibration testing is able to detect the delamination by the decreasing response of resonant frequency, however, a large variability was produced in the data taken during the summer season. This is likely due to the large variations of asphalt temperature, consequently changing the asphalt properties. For the specific case, machine learning is used to assist in capturing the debonding condition of the layers. The experimental data is first studied by supervised learning using a support vector machine (SVM); however, the damage was not detected well due to the lack of specific examples (i.e., delamination of the layers with exact depth) for input. Instead, machine learning using a principal component analysis (PCA) is applied where only standard data (i.e., without delamination) is given so that when an abnormality such as delamination of the layers is present, it is detected. Using this method, the presence of the delamination of the pavement layers was identified by an accuracy rate of around 90%. This indicates the proposed method, based on local vibration testing, can detect delamination of asphalt pavements with a high accuracy.

Keywords Asphalt pavement · Machine learning · Damage detection · Vibration testing

Y. Fujisaku · H. Naito (✉)
Department of Civil and Environmental Engineering, Tohoku University, Sendai, Japan
e-mail: hideki.naito.c2@tohoku.ac.jp

Y. Shirai
Research Institute, NIPPO Corporation, Kusatsu, Japan

T. Maeshima
Department of Civil Engineering, College of Engineering, Nihon University, Chiyoda City, Japan

S. Ichimaru · J. E. Bolander
Department of Civil and Environmental Engineering, University of California, Davis, US

M. Abdel Wahab (ed.), *Proceedings of the 2nd International Conference on Structural Damage Modelling and Assessment*, Lecture Notes in Civil Engineering 204,
https://doi.org/10.1007/978-981-16-7216-3_8

1 Introduction

For the pavement of Japanese airfield runways, asphalt pavements are often used. Repairing asphalt pavements is easier than for concrete pavements, however, asphalt pavements typically lack durability. Asphalt pavements have several layers, and each layer is made of different materials. The main cause of the damage of asphalt pavements is due to delamination between layers. In many airfield runways, damages due to delamination between layers has been found and accompanied by, for example, blistering, and separation of asphalt and aggregate. Therefore, it is important to detect the delamination before the damage comes to be visible on the surface of pavements.

Several studies have been conducted to detect delamination between layers of airfield runways. Cook et al. used strain gauges installed between layers of airfield pavement in use to detect delamination [1]. Chun et al. evaluated the effect of different conditions on bond between layers using finite element analysis and full-scale field tests [2]. These studies focus on vehicle loads, and strain gauges need to be installed between layers of pavements, so more convenient ways are desirable. Vyas et al. applied infrared thermography to detect debonding condition [3], but thermographic methods largely depend on temperature.

In this study, local vibration tests were conducted on a modeled airfield runway. Frequency responses can be measured with a local vibration test. The resonant frequency of a damaged structure is less than that of the corresponding intact structure. Naito et al. applied a local vibration test for RC members, and found local vibration testing is effective for damage detection [4]. However, only the information of resonant frequencies was used in that study, so amplitude spectrum that contains various modes has not been utilized so far. Herein, we applied machine learning, which uses the amplitude spectrum for delamination detection, so that more information can be included into detection.

The objective of this study is to assess whether select machine learning methods are useful for the delamination detection of multi-layer asphalt pavement. As part of this research, model asphalt pavement of airfield runways was constructed. Local vibration tests were conducted, and we measured amplitude spectrum. From the amplitude spectrum, we conducted delamination detection with two machine learning methods: support vector machine [5] and principal component analysis-based algorithm [6].

2 Experimental Program

2.1 Asphalt Pavement Field

A pavement that imitated actual airfield runways was installed, as shown in Fig. 1. This pavement has two regions that differ in the thickness of layers. One has thickness of 240 mm (H240 pavement), and the other has thickness of 300 mm (H300

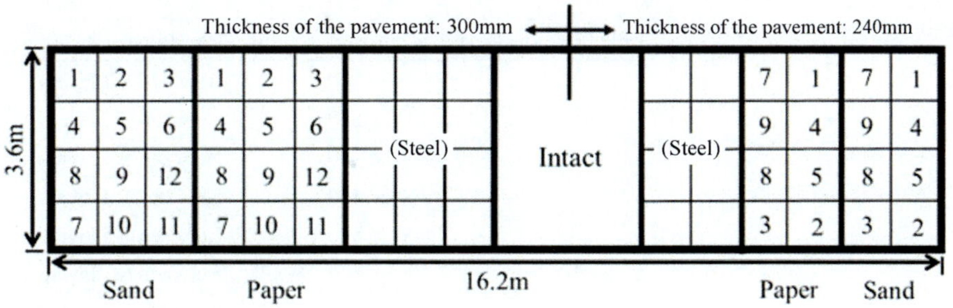

Fig. 1 Dimensions of the airfield pavement specimen (Numbers indicate types of delamination in Table 1)

pavement). Each of them has a surface layer, an intermediate layer and a base layer. Each layer has the same thickness. The surface layer is made of a dense particulate asphalt mixture. The intermediate one is made of a coarse particulate asphalt pavement. The base one is made of a large particulate asphalt mixture. The mass densities as design values are 2,350 kg/m^3, 2,350 kg/m^3, and 2,400 kg/m^3 respectively. The lower subbase is made of crushed stones.

Delamination between layers is modeled by placing sand or paper between the layers. Delamination modelled by paper simulates debonding due to lack or deterioration of asphalt emulsion between layers, and the one modelled by sand simulates a chemical change of the asphalt mixture due to stagnation of moisture between layers. They consist of two paper sheets and sand layer with a thickness of 12 mm, respectively. The locations and sizes of the modeled delamination are shown in Table 1. We also installed steel plates, as shown in Fig. 1, but they were for another examination, so measurements of areas that have steel plates are not included in this study. Sectional plans are shown in Fig. 2. Figure 3 shows the installation of modeled delaminations.

2.2 *Local Vibration Tests*

The local vibration tests were conducted on the surface of the pavement, as shown in Fig. 4. The tests were held in the summer and winter because the dynamic elastic modulus of asphalt pavement varies depending on its temperature [7]. In the tests in the summer (Aug. 2019), the vibrator was controlled: the power spectral density was 0.1(m/s^2)^2/Hz, and the frequency range was 300–20,000 Hz. The temperature of the pavement was 40 ~ 50°C. Tests were conducted 30 times on the intact points and once on each anomalous point of H240 and H300 pavement, respectively. In the tests in the winter (Dec. 2020), the vibrator was controlled: the power spectral density was 0.5(m/s^2)^2/Hz, and the frequency range was 500–20,000 Hz. The temperature of the pavement was 4.5 ~ 18.3°C. Tests were conducted 150 times on the intact points and 5 times on each anomalous point of H240 and H300 pavement, respectively.

Table 1 Delamination type

Delamination Type	Sizes (mm × mm)	Locations
1	100 × 100	Between S-I layers
2	200 × 200	Between S-I layers
3	300 × 300	Between S-I layers
4	400 × 400	Between S-I layers
5	500 × 500	Between S-I layers
6	600 × 600	Between S-I layers
7	200 × 200	Between I-B layers
8	400 × 400	Between I-B layers
9	600 × 600	Between I-B layers
10	100 × 100	Between S-I layers
	200 × 200	Between I-B layers
11	200 × 200	Between S-I layers
	400 × 400	Between I-B layers
12	300 × 300	Between S-I layers
	600 × 600	Between I-B layers

*Note*Between S-I layers = Between surface and intermediate layers. Between I-B layers = Between intermediate and base layers.

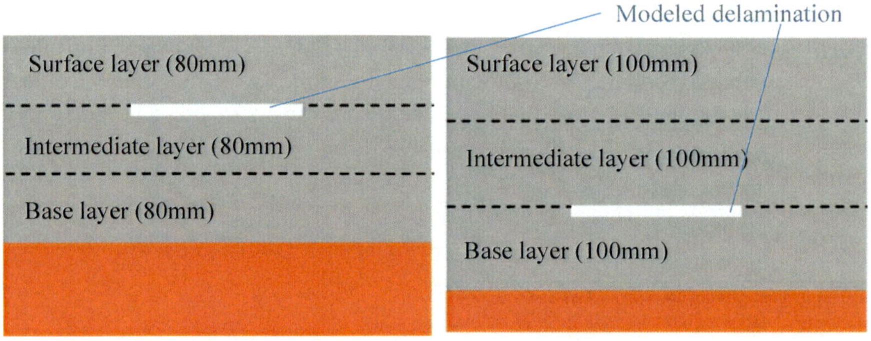

(a) H240 pavement (b) H300 pavement

Fig. 2 Sectional plans of the pavement

Figures 5, 6, 7 show the frequency responses of three points on H240 pavement, which are an intact point, and the points which have delamination type 2 and 5 modeled by sand layers, respectively. In the measurements of anomalous points, the frequency responses vary from that of the intact point because of diffraction and reflection around the delamination [4, 8], as shown in Fig. 4.

In this study, resonant frequencies were decided from the following steps. First, the calculated resonant frequency f_{cal} is

Fig. 3 Installation of modeled delaminations

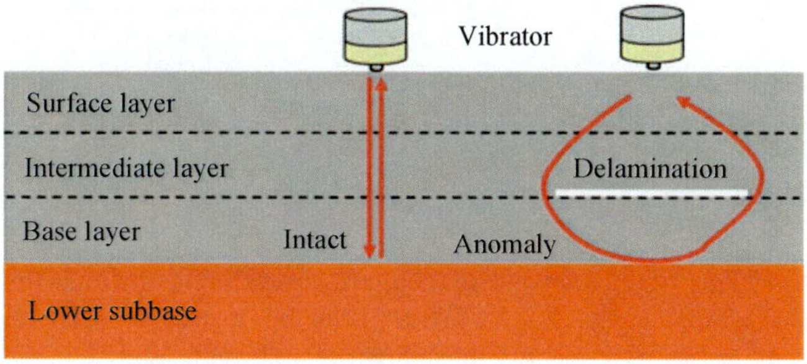

Fig. 4 Application of vibration testing for asphalt pavement

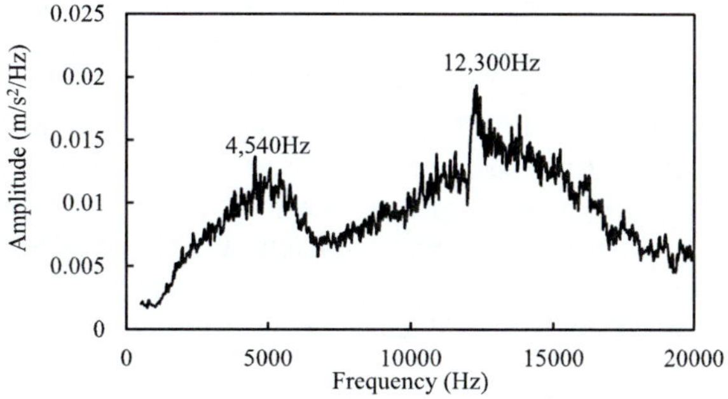

Fig. 5 Frequency responses of an intact point (H240 pavement)

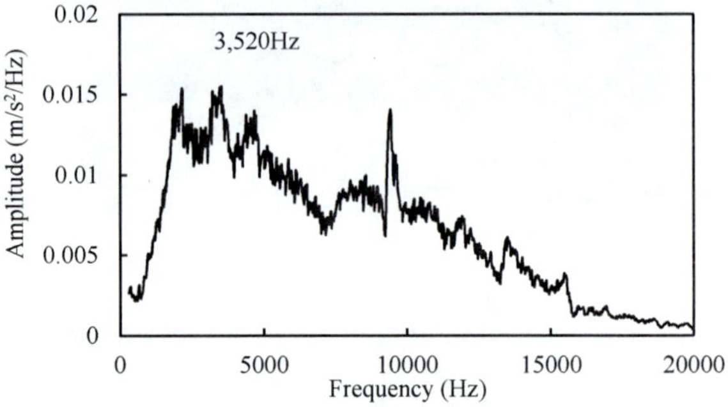

Fig. 6 Frequency responses of delamination 2 modeled by a sand layer

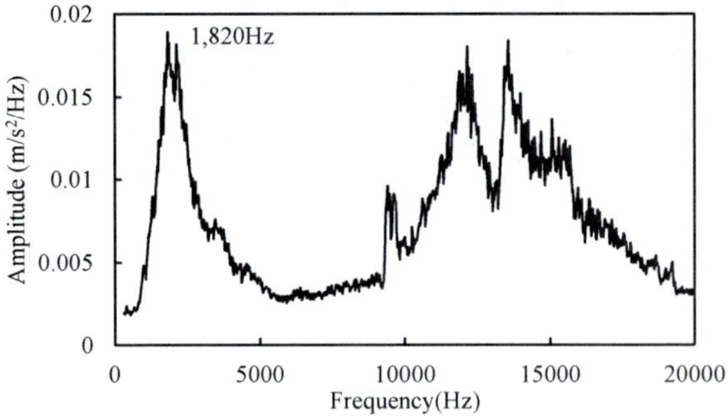

Fig. 7 Frequency responses of delamination 5 modeled by a sand layer

$$f_{cal} = \frac{c}{2H} \tag{1}$$

where c is the velocity of elastic wave in the asphalt pavement, and H is the thickness of the whole pavement layer system. From ultrasonic testing on samples acquired from the pavement, we can assume the velocities of elastic waves in summer and winter are 1200 m/s and 3000 m/s, respectively. Suppose then that the frequency that has the maximum value of amplitude in the vicinity of f_{cal} is f_{exp}, and that the mean of f_{exp} is f_m. Finally, the defined resonant frequency is the frequency which has the maximum value of amplitude in $0 \sim 1.2 f_m$(Hz). This procedure is conducted so that we can eliminate the higher modes.

Statistical analysis of the first resonant frequencies of the standard data is shown in Table 2. In Table 2, the velocities of elastic waves are calculated from the resonant frequencies and Eq. (1) and the thickness of the pavements. High coefficient of variation was obtained for the winter H300 pavement measurements. It is assumed that steel plates, which were installed between surface and intermediate layers near the intact zone, had influence on the standard data.

Table 2 Statistical analysis of standard data

Season	Winter (Dec. 2020)		Summer (Aug. 2019)	
Type of pavement	H240	H300	H240	H300
Temperature (°C)	4.5–18.3		40–50	
Number of data	150	150	30	30
Resonant frequency (Hz)	5080	3360	2410	2260
Standard deviation (Hz)	680	970	490	452
Coefficient of variation (%)	13.5	29.0	20.4	20.0
Velocity of wave (m/s)	2440	2020	1160	1360

3 Machine Learning Methods

We examined two approaches of machine learning methods. One is SVM (Support Vector Machine), which is known for its high accuracy, and the other is PCA (Principal Component Analysis)-based algorithm. Both algorithms were implemented in the scikit-learn [9] package. We used min–max scaling, which is also called normalization and changes the maximum value to 1 and the minimum value to 0, as a feature scaling method for amplitude spectrum, which is measured in the vibration tests. The scaled value of amplitude of each frequency was put into the machine learning methods as features.

3.1 Support Vector Machine

SVM is a machine learning method mainly for binary classification. We use a radial basis function (RBF) as its kernel function. Its main parameters are cost parameter and gamma parameter, which are set to 100 and 0.01, respectively. We conducted classifications on several conditions. First, the method of leave-one-out cross-validation (LOOCV) was conducted. In this method, learning machine trains all data but a datum, and predicts the class of that datum. This procedure continues until predictions of all data are made. Suppose that the trial with LOOCV is Trial-LOOCV. Next, we conducted classification with limited training dataset. Because the variation of asphalt property depends on temperature, it is anticipated that we cannot get training data that is measured under same temperature as that of an inspection. Therefore, we conducted trials which only data measured in the summer was input into SVM, for predicting the class of data measured in the winter, and vice versa. Suppose the trials are Trial-S and Trial-W, respectively.

The results are shown in Tables 3 and 4. SVM can predict the state of the pavement accurately in Trial-LOOCV. However, it cannot conduct anomaly detection in Trial-S and W. It appears that SVM cannot predict the state of asphalt pavement well when there is not any dataset that is measured in the same condition as the test dataset in

Table 3 Results of Trial-LOOCV

Pavement	H240		H300	
Season	Winter	Summer	Winter	Summer
Accuracy rate (%)	99.6	93.5	91.1	92.5

Table 4 Results of Trial-S and W

	Pavement	Type	Precision	Recall	F_1 score	Accuracy rate (%)
Trial-S	H240	Intact	0	0	0	32
		Anomaly	0.33	0.91	0.48	
	H300	Intact	0.40	0.01	0.03	44
		Anomaly	0.44	0.97	0.61	
Trial-W	H240	Intact	0.40	0.07	0.11	33
		Anomaly	0.32	0.81	0.46	
	H300	Intact	0.82	0.30	0.44	57
		Anomaly	0.50	0.91	0.65	

training dataset. It is because the states of asphalt pavements vary largely according to temperature. Thus, it is thought that the use of SVM without training datasets that are measured in a same condition as test datasets is inappropriate. From this finding, SVM needs the accumulation of data under various conditions to function well in anomaly detections of asphalt pavements.

3.2 Principal Component Analysis-Based Algorithm

PCA is mainly used for dimension reduction. PCA can reduce data to a given dimension, and the reduced data can also be reconstructed. It is anticipated that if PCA is trained using only data of intact areas, PCA can reduce and reconstruct the data of intact areas well but cannot do so for the data of anomalous areas. There are differences between reconstructions of a standard datum of an intact area and that of an anomalous area, as shown in Figs. 8 and 9.

In this study, the reconstruction error $E^{(i)}$ of a datum $x^{(i)}$ which has D input variables is

$$E^{(i)} = \sum_{j=1}^{D} \left(x_j^{(i)} - \hat{x}_j^{(i)} \right)^2 \tag{2}$$

where $x_j^{(i)}$ is the j-th variable of $x^{(i)}$, and $\hat{x}_j^{(i)}$ is the j-th variable of the reconstructed datum. In this study, we reduce the data to the number of components such that the

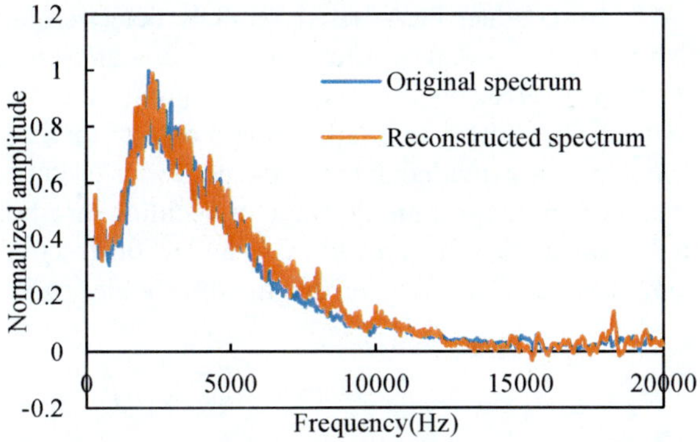

Fig. 8 Frequency responses of a standard datum and its reconstruction

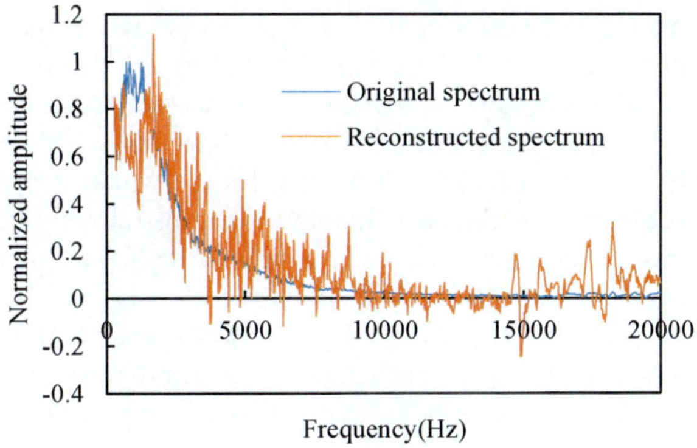

Fig. 9 Frequency responses of an anomalous datum and its reconstruction

amount of variance that needs to be explained is greater than 95%, and reconstruct them.

For the evaluation of performance, we use a Receiver Operating Characteristic curve (ROC curve). ROC curves are plotted with true positive rate as the vertical axis and false positive rate as the horizontal axis. The area under ROC curve is called the Area Under Curve (AUC), and the score of AUC is used for evaluation. Larger AUC scores indicate better performance of the PCA. The ROC curve is affected by unbalance of datasets; therefore, we conduct under-sampling for the evaluation by ROC curve. The number of data used in describing ROC curve is shown in Table 5.

Table 5 Number of data set entries

		Summer	Winter
H240 pavement	Training data	10	75
	Test data	32	150
H300 pavement	Training data	10	75
	Test data	40	150

We also conduct PCA-based anomaly detection and evaluation by LOOCV. In this LOOCV, if a datum of an intact area is chosen for a test datum, all standard data of intact areas but that datum are training data. If a datum of an anomalous area is chosen for a test datum, all standard data of intact areas are training data. Then, training data are divided in half. One portion is used to decide a thresholder, and the other one is used for training. Data for deciding thresholder are not used for training, but they are applied for trained PCA, and reconstruction errors of them are obtained. From those reconstruction errors, the thresholder th is

$$th = \mu + \sigma \tag{3}$$

where the mean value of the reconstruction errors is μ and the standard deviation is σ. If the reconstruction error of a test datum is greater than the thresholder, the datum is regarded as that of an anomalous area. If the reconstruction error of a test datum is equal or less than the thresholder, the datum is regarded as that of an intact area. The procedure of the anomaly detection algorithm is shown in Fig. 10.

The ROC curves of PCA-based algorithm are shown in Figs. 11 and 12. Each AUC is shown in Table 6. Except for the winter measurement of H300 pavement, this algorithm works well in making reconstruction error differences between intact and anomalous data. The malfunction in H300 pavement winter measurement seems to be caused by the above-mentioned reason in Sect. 2.2. The result of LOOCV with thresholder is shown in Table 7. In H240 pavement, it appears that we can conduct anomaly detection on asphalt pavements with an accuracy rate of around 90%. From Table 2, measurements in summer have low measurement accuracies; nevertheless,

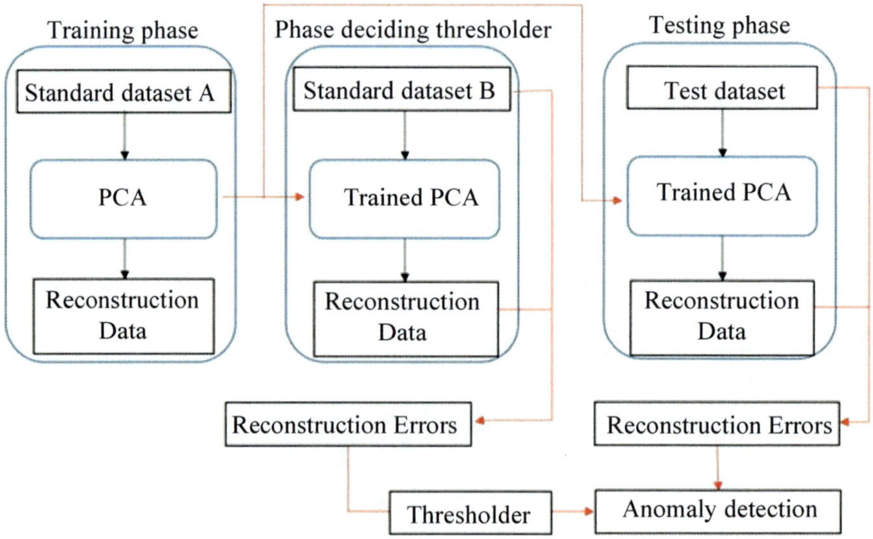

Fig. 10 The flow of PCA-based anomaly detection

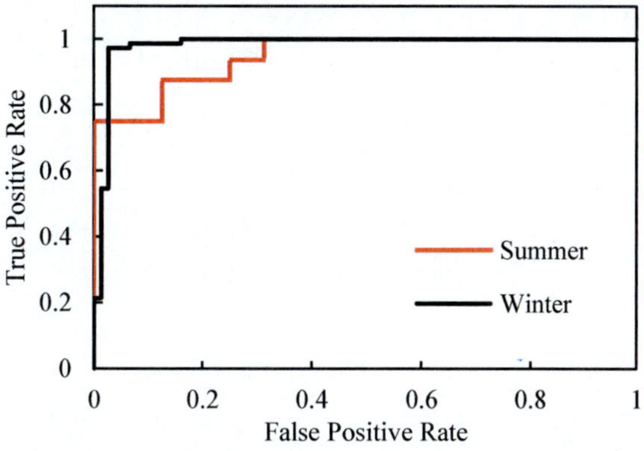

Fig. 11 ROC curve of H240 pavement

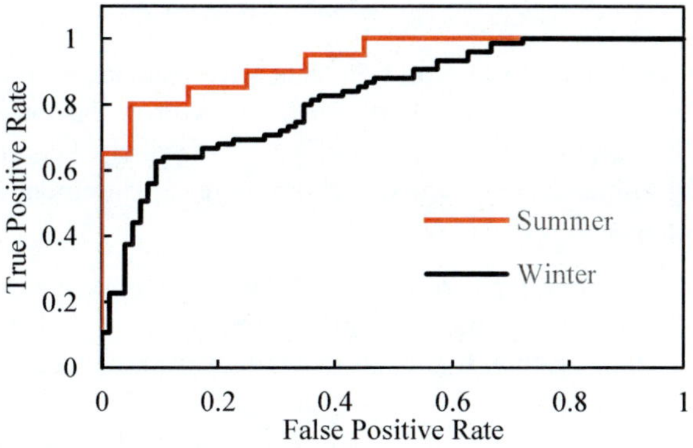

Fig. 12 ROC curve of H300 pavement

Table 6 AUC of each season and pavement

	Summer	Winter
H240 pavement	0.95	0.98
H300 pavement	0.93	0.82

the accuracy rate of H300 pavement measurement in summer is still 79%. However, for H300 pavement measurement in winter, the algorithm does not conduct anomaly detection well, as seen from the accuracy rate of 73% and low AUC value. Data measured in intact areas may have been affected by the presence of the steel plates and contains many incorrect data as seen from the high coefficient of variation value in Table 2.

In practical use, once we detect standard areas of a pavement, we can conduct anomaly detection on the whole pavement with the PCA-based algorithm from the obtained standard data. This method enables large time savings, since it is not necessary to check all obtained data visually. The determination of thresholder value is still an issue, but this method functions satisfyingly so far.

Table 7 Result of LOOCV

Pavement	Season	Type	Precision	Recall	F_1 score	Accuracy rate (%)
H240	Summer	Intact	0.93	0.90	0.92	89
		Anomaly	0.82	0.88	0.85	
	Winter	Intact	0.97	0.97	0.97	96
		Anomaly	0.94	0.94	0.94	
H300	Summer	Intact	0.79	0.87	0.83	79
		Anomaly	0.80	0.70	0.74	
	Winter	Intact	0.71	0.87	0.78	73
		Anomaly	0.77	0.55	0.64	

4 Conclusions

In this study, local vibration tests were conducted on an asphalt pavement that has modeled delamination between its layers. From the obtained data, anomaly detections with support vector machine (SVM) and principal component analysis (PCA)-based algorithms were conducted. From these measurement and detections, the following conclusions can be made:

1. The frequency responses of asphalt pavements were measured with local vibration tests. The resonant frequencies of asphalt pavements vary according to temperature. It appears that the measurement accuracy in summer is lower than that of winter.
2. SVM is an excellent method of machine learning, but without training data that are measured in similar conditions as the test data, it does not function well. However, if the accumulation of various types of data under general conditions becomes sufficient, SVM can be useful for anomaly detections of asphalt pavements.
3. The PCA-based algorithm can conduct anomaly detections satisfyingly only with standard training dataset even although the accuracy rate of the PCA-based algorithm is less than that of SVM with LOOCV. The significance of the PCA-based algorithm is to enable anomaly detections without anomalous training data, unlike other supervised learning methods.

From these findings, the anomaly detection method with local vibration testing and the PCA-based algorithm is useful for detecting delamination of asphalt pavements in airfield runways.

References

1. Cook K, Garg N, Singh A, Flynn M (2016) Detection of delamination in the HMA layer of runway pavement structure using asphalt strain gauges. J Transp Eng 142(11):04016047
2. Chun S, Kim K, Greene J, Choubane B (2015) Evaluation of interlayer bonding condition on structural response characteristics of asphalt pavement using finite element analysis and full-scale field tests. Constr Build Mater 96:307–318
3. Vyas V, Patil VJ, Singh AP, Srivastava A (2019) Application of infrared thermography for debonding detection in asphalt pavements. J Civ Struct Heal Monit 9(3):325–337
4. Naito H, Bolander JE (2019) Damage detection method for RC members using local vibration testing. Eng Struct 178:361–374
5. Vapnik VN (1999) An overview of statistical learning theory. IEEE Trans Neural Netw 10(5):988–999
6. Abdi H, Williams LJ (2010) Principal component analysis. Wiley Interdiscip Rev: Comput Stat 2:433–459
7. Ryden N (2011) Resonant frequency testing of cylindrical asphalt samples. Eur J Environ Civ Eng 15:587–600
8. Naito H, Sugiyama R, Bolander JE (2020) Local vibration testing and damage evaluation for RC bridge decks. J Struct Eng 146(9):04020168
9. Pedregosa F, Varoquaux G, Gramfort A, Michel V, Thirion B, Grisel O, Blondel M, Prettenhofer P, Weiss R, Dubourg V, Vanderplas J, Passos A, Cournapeau D, Brucher M, Perrot M, Duchesnay E (2011) Scikit-learn: Machine learning in Python. J Mach Learn Res 12:2825–2830

Damage Evaluation of RC Structures Using Simplified Wave Propagation Analysis and Machine Learning

Ryu Hashimoto, Hideki Naito, Sonoko Ichimaru, and John E. Bolander

Abstract Local vibration testing can be used to identify cracks within reinforced concrete (RC) structures; however, achieving high accuracy in the damage evaluation is a challenge. The amount of wave propagation across a crack is affected by several factors such as aggregate contact at the crack face. This study proposes a crack model that simplifies those factors in the wave propagation analysis based on the finite-difference time-domain (FDTD) method. The simplification is achieved by blocking the wave propagation across macro-crack that have a width larger than approximately 0.1 mm. The proposed crack model is validated by comparing numerical analysis and experimental results. Furthermore, a machine learning classifier is applied to the experimental and analytical data to estimate the degree of damage in the RC beams. The analytical resonant frequencies show good agreement with the experimental results of local through-thickness vibration tests on the damaged RC beam specimens. In addition to the analysis, the cracks in the RC beams are well detected by machine learning. This study shows that the proposed crack model is effective for crack identification in local vibration testing. Furthermore, machine learning contributed to improving the accuracy of damage detection.

Keywords RC beams · Concrete cracks · Non-destructive testing · Wave propagation analysis · Machine learning

1 Introduction

Deterioration of reinforced concrete (RC) structure can be determined through inspection and analysis of cracks on the concrete surface. The cracking and damage

---

R. Hashimoto (✉) · H. Naito
Department of Civil and Environmental Engineering, Tohoku University, Sendai, Japan
e-mail: ryu.hashimoto.p1@dc.tohoku.ac.jp

H. Naito
e-mail: hideki.naito.c2@tohoku.ac.jp

S. Ichimaru · J. E. Bolander
Department of Civil and Environmental Engineering, University of California, Davis, USA

M. Abdel Wahab (ed.), *Proceedings of the 2nd International Conference on Structural Damage Modelling and Assessment*, Lecture Notes in Civil Engineering 204,
https://doi.org/10.1007/978-981-16-7216-3_9

condition of buildings and infrastructure provides resources for making maintenance decisions in order to keep the various standards, such those related to safety and living comfort of the structures.

However, the evaluation of the damage may require much effort, for instance, when the crack does not appear on the surface which is difficult to be captured by inspection, or when the concrete surface is covered by a material such as carbon fiber sheet. In those situations, various non-destructive testing (NDT) methods (e.g., ultrasonic method [1, 2], impact elastic wave method [3], thermography [4], electromagnetic wave radar [5, 6]) are effective means for evaluating the deterioration of the structures. A number of existing non-destructive methods are able to capture the deformation of the RC structure, however, those are often limited to evaluating the damage near the surface.

On the other hand, the vibration test method that generates a longer wavelength is able to estimate the damage inside concrete as diffraction and reflection of the waves by aggregate or rebars can be suppressed. Naito et al. identified cracking in concrete structures by studying the response of the resonant frequency in a local through-thickness vibration test [7–10]. The study suggests the use of a small size vibrator that locally excites a structural member.

Based on the method proposed by the authors, this study further examines the application scope of the local vibration test and its accuracy through numerical simulation and machine learning. The simulation, which can visualize the wave propagation behavior around cracks, is expected to facilitate the understanding of the wave propagation mechanism within damaged concrete. Simplification of the wave propagation is applied in the analysis using the finite-difference time-domain (FDTD) method. In particular, the particle velocities vanish at macro-cracks larger than 0.1 mm in width [7]. The validity and advantage of the proposed crack model are evaluated by comparing the analytical results with the experimental results of the vibration test performed on a damaged RC beam specimen. The damage in the RC structure is diagnosed through the response of the resonant frequency in the local vibration test as well as damage classification by machine learning using the frequency response as sample data.

2 Damage Evaluation by Wave Propagation Analysis and Local Vibration Testing

2.1 Experimental Program

Cracks in RC beams are identified using local vibration tests [7, 8]. Twelve RC beams were loaded in flexure by several loading steps. Among the beams, nine are designed for flexural failure, and the rest for shear failure. Table 1 shows the specifications and excitation frequency band of the twelve specimens tested. Figure 1 shows the

Table 1 Specifications and excitation frequency band of RC beams

Specimen	Size			Concrete property		Excitation frequency band
	Length	Height	Width	Speed of sound	Mass density	
	(mm)	(mm)	(mm)	(m/s)	(kg/m^3)	(Hz)
F560-1.6 [7]	1800	400	200	3720	2300	1000 ~ 10,000
F560-1.9 [7]	1800	350	200	3720	2300	1000 ~ 10,000
F560-2.2 [7]	1800	300	200	3720	2300	1000 ~ 10,000
F560-2.8 [7]	1800	250	200	3720	2300	1000 ~ 10,000
F700-2.0 [7]	1800	400	200	3720	2300	1000 ~ 10,000
F700-2.3 [7]	1800	350	200	3720	2300	1000 ~ 10,000
F700-2.8 [7]	1800	300	200	3720	2300	1000 ~ 10,000
F700-3.5 [7]	1800	250	200	3720	2300	1000 ~ 10,000
F900-3.9 [8]	2400	300	600	3610	2290	1000 ~ 8000
S350-1.0	1200	400	400	3730	2330	500 ~ 5000
S900-3.9 [8]	2400	300	600	3610	2290	1000 ~ 8000
S2000-2.8	5000	800	800	3540	2250	2000 ~ 3000

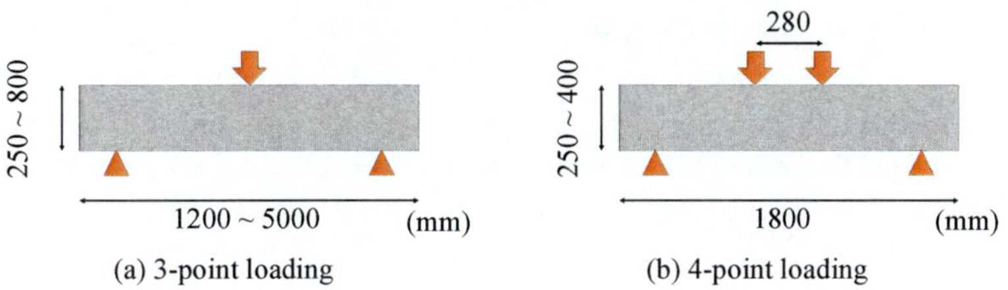

Fig. 1 Schematic diagram of RC beam specimen under (**a**) 3-point loading and (**b**) 4-point loading

schematic images of the loading. Specimen names in Table 1 indicate the type of failure (i.e., flexure or shear), span length (mm), and shear span ratio.

Flexural beams were loaded until the tensile reinforcing bar reached yield followed by load removal. A vibration test was conducted before any loading step and a crack diagram was produced after unloading. Those were repeated until the displacement reached several multiples of the yield displacement (i.e., δy). On the other hand, loading was terminated for shear beams when the shear cracks occurred and the load started to decrease. Similar to the flexural beams, a vibration test was conducted on shear beams and a crack diagram was produced after each load removal.

A schematic diagram of the vibration tests (i.e., local through-thickness and longitudinal) is shown in Fig. 2. The vibrator locally excites the member in the transverse direction through the generation of frequency sweep or white noise excitation in the local through-thickness vibration test. Two sensor locations are sought in this study, either on the opposite face or on the same surface as the vibrator. While this study presents only the results of the former sensor configuration, the latter takes into

(a) Local through- thickness vibration test (b) Longitudinal vibration test

Fig. 2 Schematic diagram of (**a**) local through-thickness vibration test and (**b**) longitudinal vibration test

account the typical test condition on pavements where the vibrator and the sensor are located on the same surface. According to the measured frequency, the decrease in the resonant frequency can be used to estimate damage within concrete. Specifically for the F560 series and F700 series specimens, the vibration test was performed only after the displacement reached 6δy.

The longitudinal vibration test shown in Fig. 2b is conducted on F700 series specimens and F900 series specimens. This test is expected to evaluate the average rigidity of the members. As shown in the figure, the vibrator is applied horizontally from the end face of the specimen.

2.2 *Overview of Wave Propagation Analysis*

Wave propagation analysis was performed on the RC beam specimens using the two-dimensional FDTD method. The wave propagation is expressed by the following equations [9].

$$v_x^{n+\frac{1}{2}}\left(i+\frac{1}{2}, j\right) = v_x^{n\frac{1}{2}}\left(i+\frac{1}{2}, j\right)\frac{\Delta t}{\rho \Delta x}\{p^n(i+1, j)p^n(i, j)\} \tag{1}$$

$$v_y^{n+\frac{1}{2}}\left(i, j+\frac{1}{2}\right) = v_y^{n\frac{1}{2}}\left(i, j+\frac{1}{2}\right)\frac{\Delta t}{\rho \Delta y}\{p^n(i, j+1)p^n(i, j)\} \tag{2}$$

$$\begin{aligned} p^{n+1}(i, j) = p^n(i, j) - \rho c^2 \frac{\Delta t}{\Delta x}\left\{v_x^{n+\frac{1}{2}}\left(i+\frac{1}{2}, j\right) v_x^{n+\frac{1}{2}}\left(i\frac{1}{2}, j\right)\right\} \\ -\rho c^2 \frac{\Delta t}{\Delta y}\left\{v_y^{n+\frac{1}{2}}\left(i, j+\frac{1}{2}\right) v_y^{n+\frac{1}{2}}\left(i, j\frac{1}{2}\right)\right\} \end{aligned} \tag{3}$$

where c is the speed of sound, ρ is the mass density, p is the sound pressure, v_x and v_y are the two components of particle velocity, Δx and Δy are the spatial discretization size, and Δt is the time discretization step. The indices i and j are the spatial points and the index n is the time point.

The mesh size Δx and Δy is 2 mm, the time discretization Δt is 0.376 μs, the number of time steps is 2^{16} for this study. The speed of sound and mass density of the concrete in the RC beam specimen are given to the material properties, c and ρ.

The sound pressure is given as a sweep sine wave so that frequency is increased with the excitation time. According to the time history response of the sound pressure obtained by the analysis, the frequency spectrum is estimated through fast Fourier transform (FFT). The resonant frequency is then defined by the largest amplitude in the frequency response. Attenuation due to wave propagation and shear waves is ignored in this study [9].

Particle velocity $v_x = v_y = 0$ is given as the boundary condition on the surface of the specimen. The velocities are also set to zero $v_x = v_y = 0$ at cracks whose width is larger than approximately 0.1 mm [7].

2.3 Result of Wave Propagation Analysis

Examination of Crack Model by Longitudinal Vibration Test. Appropriate modeling of cracks is important for visualization of wave propagation behavior and reduction of calculation cost. In this study, we assume a simple crack model and examine the reproducibility of wave propagation behavior in the experiment.

The crack diagram of the RC beam is imported as a PNG file and processed so that the cracks have a width of approximately 1 pixel. The coordinates of the cracks are obtained using OpenCV, an open-source library for computer vision [11]. Figure 3 shows an example of those results.

Here, the effect of the crack width in the analysis is determined through two different modeling approaches. The first model, Model w0, assumes that the crack width is zero, whereas the second model, Model w2, assigns a void element (2 mm in size) to the crack width. Particle velocities are set to zero at the cracks as illustrated in Fig. 4.

Figure 5 compares the frequency response of the two crack models for F700-3.5 specimen under longitudinal vibration test loaded up to 1δy. The figure shows that the two models are in good agreement regardless of the crack width model. Furthermore, the analytical models fit well with the experimental resonant frequency at each loading step for the F700-3.5 specimen (Fig. 6). Other specimens also showed

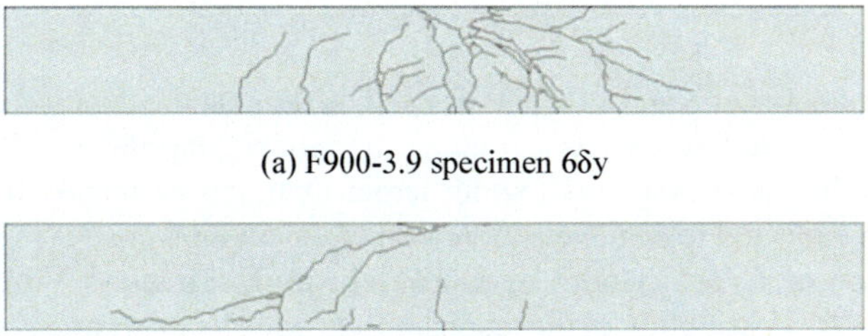

(a) F900-3.9 specimen 6δy

(b) S900-3.9 specimen 1δy

Fig. 3 Image-processed crack diagram of RC beams under flexural loading

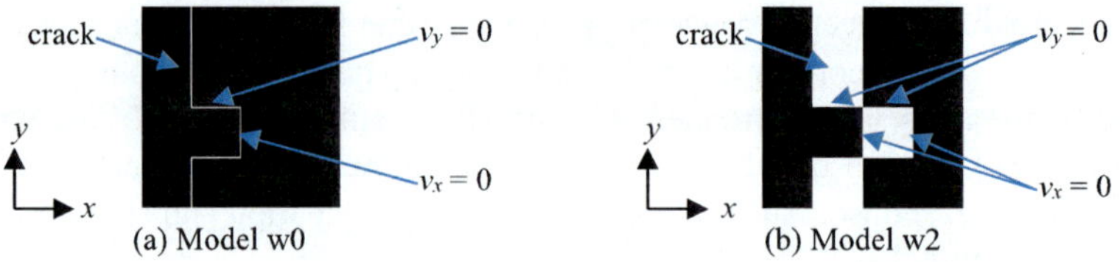

Fig. 4 Schematic images of the modeled crack width and specified boundary conditions

Fig. 5 Simulated frequency response of F700-3.5 specimen at 1δy displacement under crack model (Model w0 and w2) in longitudinal vibration test

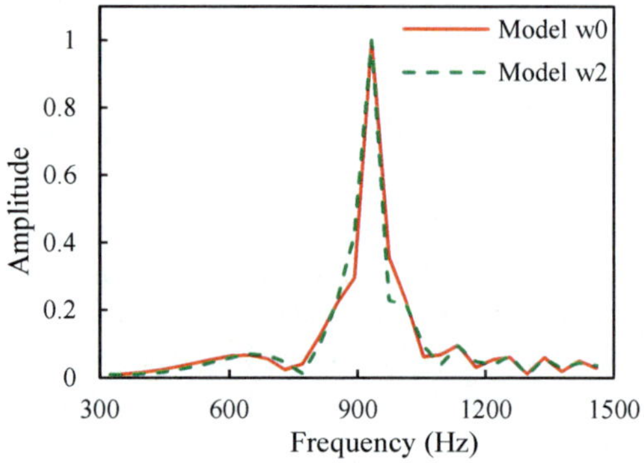

Fig. 6 Resonant frequency in longitudinal vibration test and analysis of F700-3.5 specimen

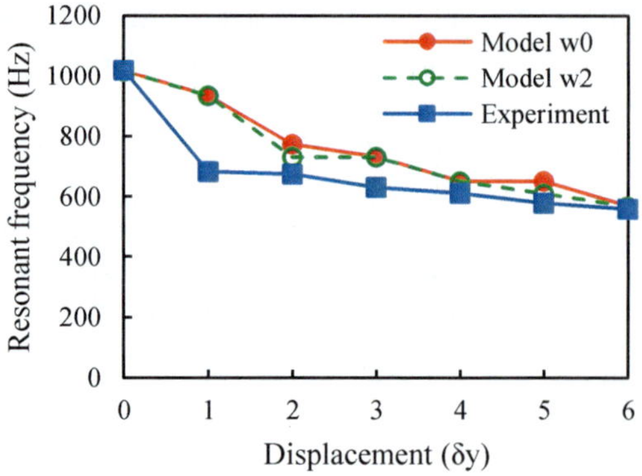

good agreement between the results thereby indicating that the detailed modeling of the crack width would have less effect under the model simplification, which gives zero particle velocities at the crack width larger than approximately 0.1 mm. The crack model is expected to achieve a reduction of calculation costs.

Examination of Crack Model by Local Through-thickness Vibration Test. Figure 7 shows the distribution of the resonant frequency by experiment and analysis for the local through-thickness vibration test in Fig. 2. Since the wave propagation across cracks is blocked under the analysis condition (i.e., zero particle velocities given for crack width larger than 0.1 mm), the sound pressure response becomes zero

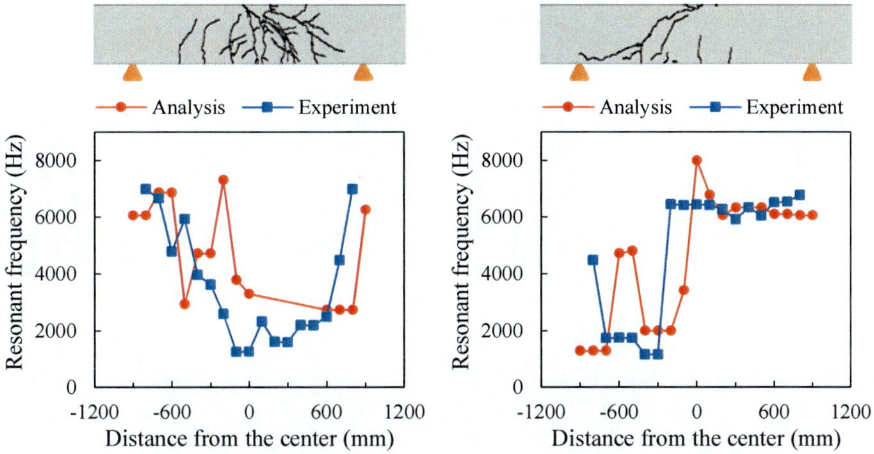

(a) F900-3.9 specimen 6δy

(b) S900-3.9 specimen 1δy

Fig. 7 Distribution of resonant frequency for local through-thickness vibration test

when the excitation point and the response measuring point are divided by the cracks. In this case, the data plot was removed from the graph. The F900-3.9 specimen in Fig. 7a has extensive flexural cracks over a wide range around the loading point. The drop of resonant frequency is seen at those locations. Similarly, the drop of the resonant frequency of the S900-3.9 specimen is noticed near the shear cracks which occurs mainly on the left side of the span. No significant decrease in resonant frequency in either the experiment or the analysis is seen on the right side, where cracking is small (Fig. 7b). The figure also shows that the analytical values of the two specimens agree well with the experimental values.

While the damages in the RC beam specimens in Fig. 7 are predicted well by the drop of resonant frequency, defined as the frequency at which the amplitude is maximum, some specimens failed to show such response. Instead, several peaks occurred without a decrease in the resonant frequency in those specimens as expressed in Fig. 8. The response seen in Fig. 8b are the result of the wave propagation affected by the diffraction or reflection around the crack.

Because of the difficulty to identify cracks by the resonant frequency, analyzing the overall behavior of the frequency response is necessary for accurate damage detection.

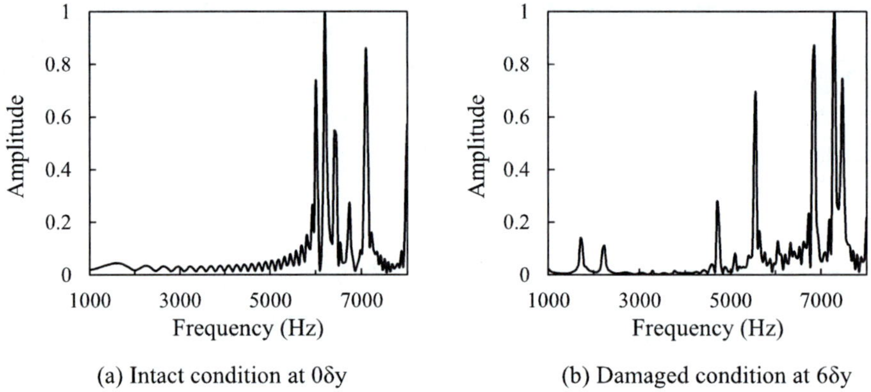

Fig. 8 Frequency response of F900-3.9 specimen at −200 mm from center

3 Damage Evaluation by Machine Learning

3.1 Overview of Analysis Method

The frequency response shown in Fig. 8 contains information such as diffraction and reflection around cracks and reflections from distant surfaces in the local through-thickness vibration test. To consider damage from the measured frequency requires effort due to the complexity of the response. Therefore, this study employs machine learning to capture the characteristic of the changes in the frequency when damage exists in the concrete, and facilitate damage identification.

A random forest classifier of scikit-learn (ver. 0.22.1) [12], which is an open-source machine learning library, is used for damage classification. The obtained frequencies from the experiment and the analysis are provided to the classifier. The number of decision trees is set to 100. The maximum amplitude of frequency response obtained in the experiment and analysis is normalized to 1. In addition, the frequency response with a smaller number of data points is linearly interpolated according to the largest frequency band, and the number of features is made consistent. The experimental and analytical dataset has 14,001 (in 0.5 Hz increments at 1000–8000 Hz) and 173 (in approximately 40 Hz increments at 1000–8000 Hz) number of features, respectively.

Two classifications are considered, intact and damaged, depending on the degree of crack propagation with respect to the height of the cross-section of the RC specimens. Figure 9 shows a schematic diagram of effective measurement range in local vibrating on the beam. Specifically, the horizontal effective measurement range including the excitation point (equivalent to cross-sectional height) of the local vibration test is defined as H [10], and the horizontal projection length of a crack within the effective measurement range is defined as D. Figure 9 shows examples of the D/H

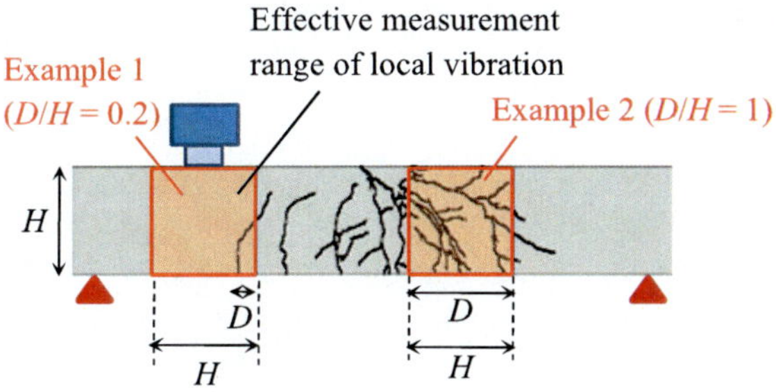

Fig. 9 Schematic diagram of effective measurement range in local vibration testing

Table 2 The number of samples used for training and testing

Criteria of *D*/*H*	Label	Analytical data	Experimental data
0.2	Intact	528	131
	Damaged	395	194
0.5	Intact	728	183
	Damaged	195	142

ratios of 0.2 and 1. The ratio D/H indicates damage level of the effective measurement range. From previous work [10], the relative dynamic elastic modulus E/E_0 is calculated from the D/H ratio; for example, 0.2, 0.5, and 1.0 in D/H correspond to 80, 55, and 4 in E/E_0, respectively. In this study, any D/H ratio that satisfies either $D/H > 0.2$ or $D/H > 0.5$ is defined as damaged.

Table 2 shows the number of samples used for each class based on the two criteria of D/H. It excludes the result of the F900-3.9 specimen at 6δy and the S900-3.9 specimen at 1δy in the training and testing phase as those specimens are performed as a blind test to examine the damage distribution explained in the following section. Since the data in Table 2 has an unbalanced distribution of the number of data available in each category, under-sampling is performed. It randomly deletes the instances of the majority class according to the number of instances of the minority class. According to the sampling, 80% of the data set was divided into the training set and 20% into the test set.

3.2 *Result of Confusion Matrix*

Table 3 shows the confusion matrix of the class prediction performed on the analytical and experimental dataset introduced in Sect. 2. The accuracy of prediction is higher for the analytical dataset than the experimental dataset as a larger number of instances is available by the analysis. Additionally, the decrease of accuracy in the experimental dataset is due to the large variability in the physical tests. The specimen boundary in the experiment is a free end, whereas the boundary condition is set to a fixed end

Table 3 Confusion matrix

Criteria of D/H	Label	Predicted label		Accuracy
		Intact	Damaged	
a. Analytical data				
0.2	Intact	72	7	0.94
	Damaged	2	77	
0.5	Intact	33	6	0.90
	Damaged	2	37	
b. Experimental data				
0.2	Intact	22	5	0.72
	Damaged	10	16	
0.5	Intact	26	3	0.81
	Damaged	8	20	

in the wave analysis. All changes in the frequency response are therefore due to an increase in crack length, which led to the increase in the accuracy of the analytical dataset. Moreover, the flexural vibration measured in the experiments is overfitted in the experimental data. This increased the misclassification in the machine learning as the frequency response changes with the decrease in stiffness.

However, even the lowest accuracy shown in Table 3 is exceeding 70% which is reasonably high. It is shown that the application of machine learning is effective for finding damage in the concrete, based on the obtained frequency response. The features of damage were found within the variety of the measured responses and the accuracy of damage prediction was high by machine learning.

3.3 *Result of Damage Distribution*

To further examine the analysis accuracy by machine learning, the probabilistic classification was performed on the F900-3.9 specimen at 6δy and the S900-3.9 specimen at 1δy. The prediction probability classified as the intact class for the RC beam specimens was calculated for the experimental measurement points for D/H = 0.2 and D/H = 0.5 criteria, respectively (Figs. 10 and 11). Here, the prediction probability as the intact class is the ratio of the decision trees which predicted to be the intact class among all decision trees in the random forest.

Results of the F900-3.9 specimen show that the prediction probability as the intact class significantly reduces at the crack locations for both analytical and experimental datasets (Figs. 10 and 11). This behavior follows the actual failure that occurred in the specimen. In addition, most of the misclassifications that occurred at measurement positions where D/H is near to the criteria are reasonable for this type of study. A similar trend is also observed in the S900-3.9 specimen. These results indicate that, in

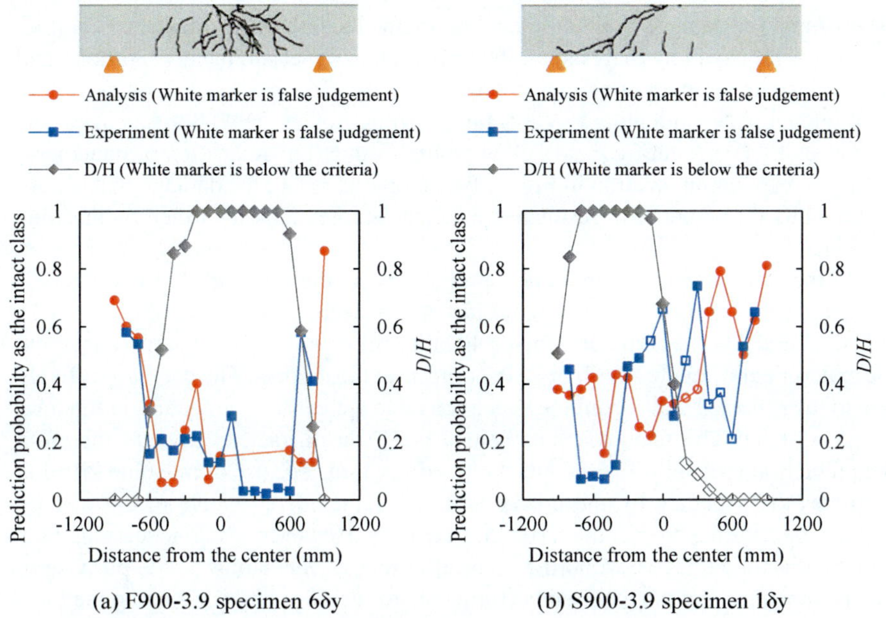

Fig. 10 Prediction probability as intact class (criteria of $D/H = 0.2$)

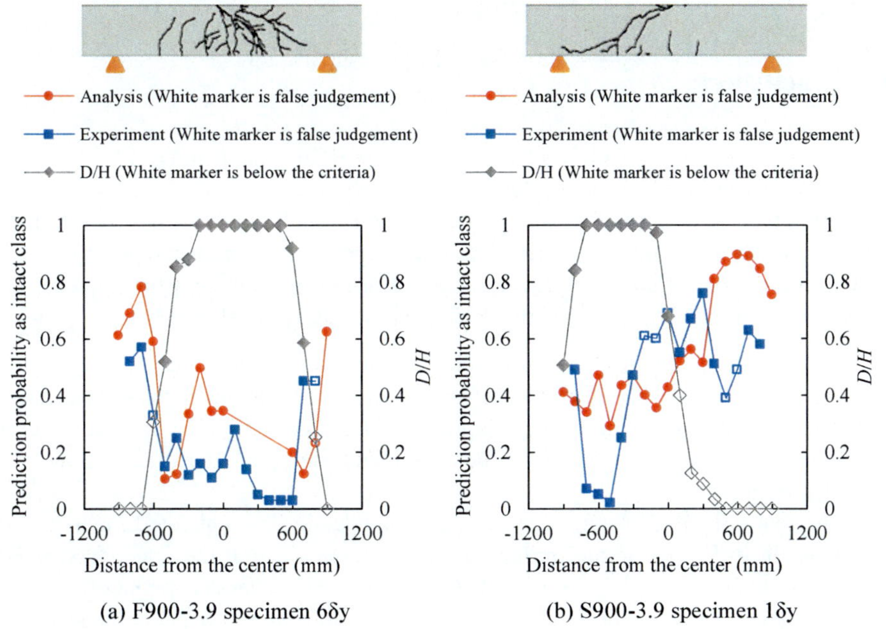

Fig. 11 Prediction probability as intact class (criteria of $D/H = 0.5$)

establishing the damage evaluation method by the local through-thickness vibration test, wave propagation analysis is effective to clarify measurement conditions and show adequate accuracy for damage detection.

While damage is identified by machine learning for the F900-3.9 specimen, the response of its resonant frequency shows a high value (i.e., indication of intactness) at the corresponding location in Fig. 7. Based on this result, the damage that cannot be found by the response of resonant frequency is potentially identified by machine learning.

On the other hand, the prediction probability as the intact class for the S900-3.9 specimen decreases on the right side of the beam where there are no cracks (Figs. 10 and 11). In particular, the probability drops in both the analysis data and the experimental data for the $D/H = 0.2$ criteria at those spots. The decrease is likely due to the reflected wave from the specimen edge appearing as a feature in the low-frequency band. This feature is thought to be due to the diffracted wave, and therefore the prediction probability as the intact class decreased. The use of machine learning increases the accuracy to find cracks within RC members relative to the method that investigates the drop of the resonant frequency. However, machine learning may pick up distant damage beyond the inspection range. Depending on the purpose of the maintenance and the required performance for damage diagnosis using the local vibration test, it is suggested to study both the response of resonant frequency and the classification by machine learning.

4 Conclusion

This study performs wave propagation analysis on RC beam specimens with flexural and/or shear cracks to study the damage in the beams using the local vibration test. The cracks in the RC beams are located by studying the responses in the measured resonant frequency. In addition to the experimental and wave propagation analysis on the damaged RC beams, machine learning is employed for improving crack identification. The findings obtained from this study are shown below.

1. Wave propagation within a damaged RC beam is simulated using a simplified model. It assumes that the wave propagation vanishes, equivalent to zero particle velocities, at cracks with a width larger than approximately 0.1 mm. Using this model, the frequency response in the experiment was simulated and cracks were identified.
2. The internal wave propagation behavior changes in a complicated manner as a result of diffraction and reflection of an elastic wave at the crack face and specimen boundary. The proposed simplified crack model is capable of reproducing the features required for crack identification and it is expected to lead to a reduction in the calculation cost.
3. The characteristics of the frequency response are studied, and the degree of cracking is estimated by machine learning. It extracted essential features of

damage from the frequency response required for damage detection. Machine learning identified cracks with higher accuracy than the method to detect damage by the drop in resonant frequency. Further development of damage diagnosis by machine learning is expected to improve the conventional damage detection methods by non-destructive tests on RC structures.

References

1. Choi P, Kim DH, Lee BH, Won MC (2016) Application of ultrasonic shear-wave tomography to identify horizontal crack or delamination in concrete pavement and bridge. Constr Build Mater 121:81–91
2. Azari H, Nazarian S, Yuan D (2014) Assessing sensitivity of impact echo and ultrasonic surface waves methods for nondestructive evaluation of concrete structures. Constr Build Mater 71:384–391
3. Zhu J, Popovics JS (2007) Imaging concrete structures using air-coupled impact-echo. J Eng Mech 133(6):628–640
4. Gu JC, Unjoh S, Naito H (2020) Detectability of delamination regions using infrared thermography in concrete members strengthened by CFRP jacketing. Compos Struct 245:11, published online 112328
5. Bungey JH (2004) Sub-surface radar testing of concrete: a review. Constr Build Mater 18(1):1–8
6. Yu T, Twumasi JO, Le V, Tang Q, D'Amico N (2017) Surface and subsurface remote sensing of concrete structures using synthetic aperture radar imaging. J Struct Eng 143(10):11, published online 04017143
7. Naito H, Bolander JE (2021) Evaluation of internal cracks in RC beams using vibration testing and wave propagation analysis. In: Proceedings of the 10th international conference on bridge maintenance, safety and management. Japan, p 7 (in printing)
8. Kon E, Naito H, Igarashi A, Suzuki M (2016) Damage identification for RC culvert based on forced vibration test. In: Proceedings of the Japan concrete institute 38(2):967–972 (written in Japanese)
9. Naito H, Bolander JE (2019) Damage detection method for RC members using local vibration testing. Eng Struct 178:361–374
10. Naito H, Sugiyama R, Bolander JE (2020) Local vibration testing and damage evaluation for RC bridge decks. J Struct Eng 146(9):9, published online 04020168
11. OpenCV Homepage. https://opencv.org, Accessed 18 March 2021
12. Scikit-learn Homepage. https://scikit-learn.org, Accessed 18 March 2021

Damage in Mechanical and Materials Engineering

Fracture and Damage Mechanisms of Slipper-Retainer Assembly in Axial Piston Machines

Gaston Haidak and **Dongyun Wang**

Abstract Axial Piston Machine is a key component in many hydraulic, agriculture, construction, aeronautic, mobile robots, and many other fields. Its reputation stems from its compactness, relative ease in controlling the effective fluid displacement per shaft revolution, and high-pressure operation. However, it is facing problems of rupture, fracture, and damage of its elements, among which are slipper and retainer. These problems cause fatigue of the entire machine and consequently loss of energy and performance. Thus, to better understand the origin and causes and predict the exaggerated deformation and the increasing deterioration of Axial Piston Machine's components, we first investigated the theoretical modeling, including mathematical and simulation modeling of the slipper-retainer interface. The related external forces and applied loads considering the normal working condition of the machine are analyzed. Also, to better assess the effect of materials, materials with specific characteristics are used for simulating the hydrostatic pressure, stain, and elastic deformation of slipper- retainer assembly. At the end of each simulation, weariness is checked for each material. The most exposed areas of slipper and retainer to the fracture and intensive fatigue are determined. Finally, to validate our simulation results, a new pump is tested for a particular time to observe wear on the surfaces; suggestions are giving for improving the durability of axial piston machines.

Keywords Axial piston machine · Efficiency loss · Fracture and damage

1 Introduction

Axial piston machines are machines well used in actuation as power conversion systems such as hydraulics, aeronautics, and many others. Therefore, its solicitation is increasingly becoming important. At the same time, it is facing many problems that are leading to its efficiency loss. Thus, the improvement of its performance from the theoretical model to the experiment. Many researchers have done many works to

G. Haidak (✉) · D. Wang
Zhejiang Normal University, Jinhua, Zhejiang 321004, China
e-mail: haidak@zjnu.edu.cn

M. Abdel Wahab (ed.), *Proceedings of the 2nd International Conference on Structural Damage Modelling and Assessment*, Lecture Notes in Civil Engineering 204,
https://doi.org/10.1007/978-981-16-7216-3_10

improve the axial piston machines' efficiency, emphasizing the lubricated interfaces such as slipper/swashplate, piston/cylinder brock, and cylinder brock/ valve plate interfaces. Impressive works based on the analysis of thermal and hydrodynamic modeling under lubrication mechanism have been done [1–5]. To respond to the effect of the contact pressure in the problem of elastic deformation should be taken into account. Several research works in fretting contact [6–10] have been done in order to reduce fatigue strength, frictions, and wear. However, some other parts and contact interfaces, like the interface between slipper and retainer, which are not well lubricated, are also significant wear and efficiency loss. The rapid evolution of fatigue between slipper and retainer is caused partially by the unlubricated contact and friction happening in their interface.

Nomenclature:	
D Slipper outer diameter [m]	F_{wK} Centrifugal force [N]
d Slipper inner diameter [m]	F_{RK} Force of piston-slipper [N]
F_{fz} Swashplate reaction force [N]	m_s Mass of slipper [kg]
F_{SK} The piston ball force [N]	P Pressure [Pa]
F_{DK} The Displacement chamber force [N]	P_G Slipper pocket pressure [Pa]
F_{TK} Viscous friction force [N]	w Pump rotation speed [rad/s]

Based on the recent works related to the improvement of power efficiency of axial piston machines, the damage and failure on slipper/retainer have not been considered. Therefore, the study of failure and wear due to these two solids' thermal and elastic deformation (slipper and retainer) have not been considered. Significantly, for well improving the efficiency and performance of a component as an axial piston, a deep study should also consider the damage and fracture of its components.

To better solve this problem in order on the one hand to cover a systematic study of the axial piston machine and to understand the mechanism of destruction and aging of the slipper/retainer assembly, it is important to analyze the fracture and damage mechanism to which the slipper and retainer are subjected.

2 Problem Approach and Materials

Due to the complexity of the movements of the slipper on the swashplate, it is naturally subjected to more external forces. The retainer's role is to retain and reduce the eccentric movements that the slipper performs on the swashplate. Thus, the retainer is designed so that it has space between the slipper and the gaps of the retainer. Given that the pump can burp at variable speeds adds to the control movement of the angle of inclination of the swashplate, so the movement of the pad of the slipper is not in principle circular [11]; which causes the small shocks between the slipper pad and the holes of the retainer. These shocks are the primary sources of the deformation of solids (slipper and retainer). Figure 1 shows the slipper/retainer assembly location in the machine and the external loads that apply to it.

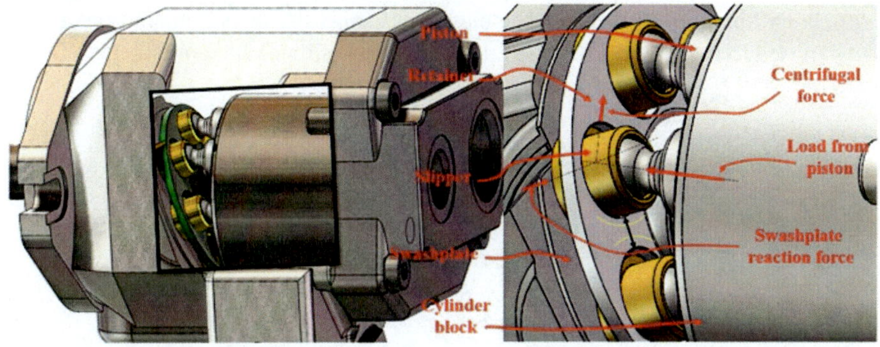

Fig. 1 View of slipper and retainer in axial piston machines

When manufacturing the pump, the ratio between the outer diameter of the slipper and the retainer hole is large (close to 1) for most of the machines is made without considering the possible exaggerated training. Thus, when the retainer holes undergo the elastic deformation (abnormal deviation), it can no longer play usually its ball of retention of the slippers. In extreme cases, the slippers can be found stuck in these holes. The slipper is pressed by more than one external force. The exaggerated deformation of the slipper comes from a part of malfunction of the pump (control problem) or the overloads. These problems can lead to the destruction of the slipper.

In most companies, slippers are designed with the pad having multiple lands. But following a malfunction and overpressure, slippers can have damage, such as can be observed in Fig. 2. After the machine's destruction in a construction company, the machine was disassembled to understand the reasons for its destruction. Thus, we observed and understood that the machine's damage came from the rupture and exaggerated formation of the slipper, as shown in Fig. 2.

Thus, to better understand this mechanism that leads to the destruction and damage of the slipper and to retain to propose suitable solutions, it is important to understand the degradation process in consideration of several parameters such as external forces and loads types of contact. The mathematical model describing this mechanism is detailed [12], and some of the applied external loads to the slipper/retainer interface are given from Eq. (1) to Eq. (4). To better understand them. Effects of the materials used to manufacture these solid axial elements of piston machine. These different

Fig. 2 Large elastic deformation, wear and fracture of the slipper and retainer

materials will be used to simulate the elastic deformation, the strain, and the areas most exposed to the fatigue of the solid components (slipper and container).

$$F_{fz} = \frac{1}{8}\pi P_G \frac{D^2 - d^2}{ln\frac{D}{d}} \tag{1}$$

$$F_{SK} = \pi \frac{\left(r_p^2 - r_0^2\right)}{8.cos\theta} P_{DC} \tag{2}$$

$$F_{TK} = \frac{\pi}{4}\left(D^2 - d^2\right)\eta\frac{\omega R}{h_G} \tag{3}$$

$$F_{\omega K} = m_G.R.\omega^2 \tag{4}$$

where m_G the mass of the slipper, ω is the angular velocity of the slipper, η is the oil viscosity, h is the film thickness, β the inclination angle of the swashplate, P_{DC} the pressure of the displacement chamber, r_p is a piston radius and r_0 the slipper orifice diameter. The weighted residuals method approximates the function for proofing the relationship between the Elastic deformation and strain, and the four-node linear tetrahedron shape is used to discretize a solid domain [12].

The relationship between stress and elastic/plastic deformation is well detailed and disassembled by the author [12] to calculate the pressure and pressure distribution in solids. Table 1 gives the characteristics of the two different materials used to simulate stress, strain, elastic deformation, and fatigue of each slipper. A new pump has been installed and tested in the test rig to see the beginning of the evolution of the slipper fatigue and grooming processes (Fig. 3).

Table 1 Materials' properties of slipper and retainer

Components	Retainer	Slipper
Proprieties/Names	**Gray Cast Iron (Sn)**	**Leaded Commercial Bronze**
Density [Kg/m^3]	*7200*	*8700*
Elastic Modulus [N/m^2]	*6.62e + 10*	*1.1e + 011*
Shear Modulus	*5e + 10*	*4.1e + 010*
Tensile Strength [N/m^2]	*1.52e + 8*	*2.55e + 8*
Thermal Cond. $W/m.K$	*45*	*26*
Specific Heat [$J/kg.K$]	*510*	*340*

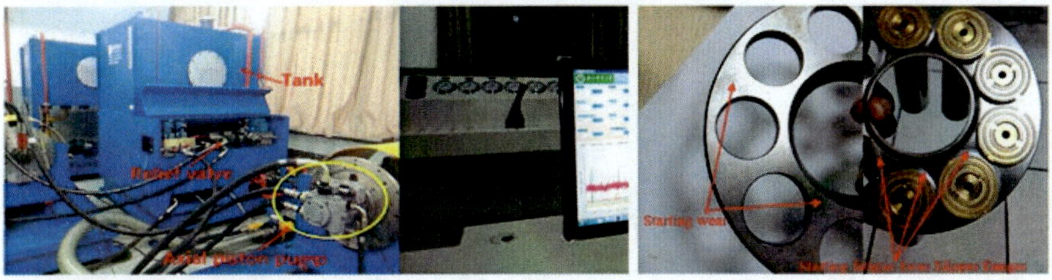

Fig. 3 Experimental environment

3 Results and Discussion

Figure 4 shows the type of mesh used for the numerical calculation of the slipper and retainer (that of the tetrahedral nodes) the pressure zone high in the retainer. This zone is also the isothermal zone. This result explains the shocks that the slippers and retainer suffer during the operation of the slipper.

The external pressures applied on the retainer are: the force of the resort, which is applied in the direction of the normal vector has a flat surface of the retainer, the pressure force of the reactionary retainer of the slipper, which applies in the same direction as the normal vector has the surface of the retainer and in the opposite direction to that of the resort, the force of contact between the slippers and the holes of the retainer (circular contact surface) and also that of the gravitation coming from the rotation of the retainer /slipper assembly.

Figure 5a presents the simulated result of the exaggerated deformation of retainer; Fig. 5b, c present respectively its related strain and fatigue. The related fatigue (areas in red color) shows zones with less safety level of the design. It also indicates the area exposed to the damage and possible fracture. For the retainer material described in Table 1, the maximum deformation reached 1.8 mm.

The simulation results of the slipper elastic deformation and fatigue are shown in Fig. 6. And we can easily notice in Fig. 6a the most compressed area of the slipper pad. The outer edge of the slipper pad is the area that undergoes the maximum

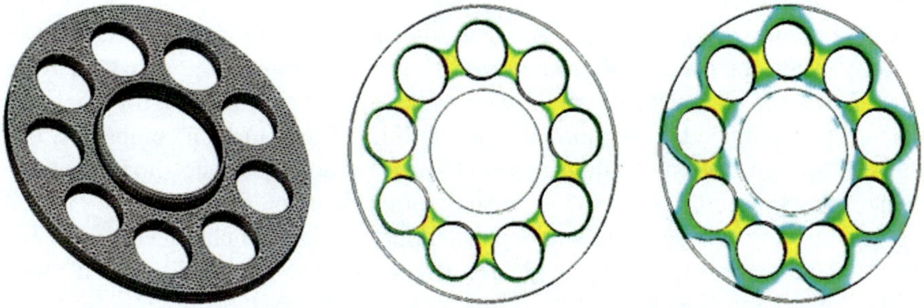

Fig. 4 Retainer mesh and iso-over loaded area

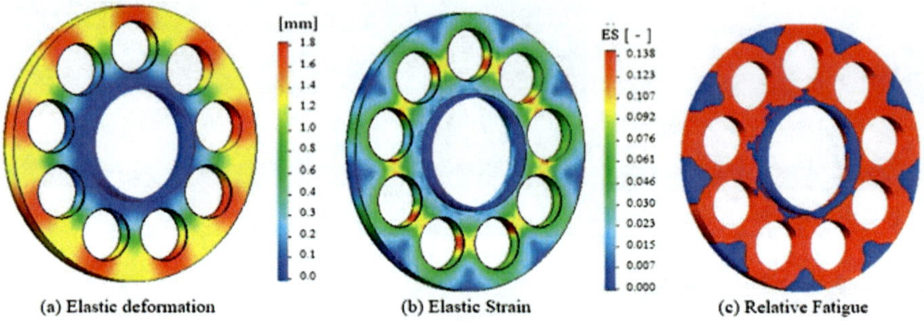

Fig. 5 Retainer's pressure, elastic deformation, strain, and fatigue

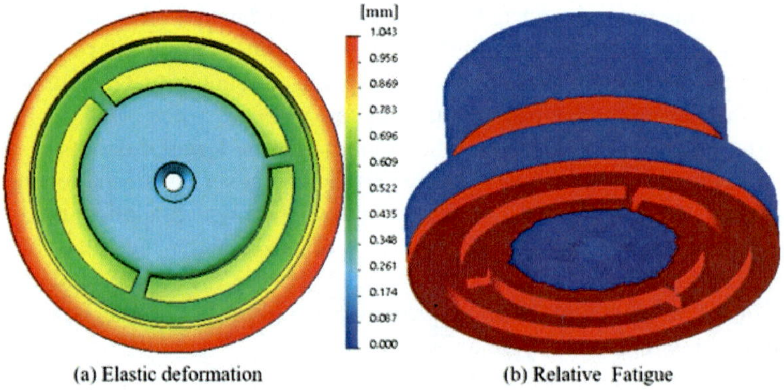

Fig. 6 Slipper's elastic deformation and fatigue

deformation (1,043 mm), and the area most exposed to destruction or damage is those whose fatigue represents (Fig. 5b). The fatigue area located on the upper part of the slipper (Fig. 6) results from the shocks and friction between the slipper and piston head. The fatigue zone located on the slipper pad is explained by the fact that the slipper does not slide circularly on the swashplate but also performs dynamic slopes during its rotation.

A new pump has been insured and test; after a length of time, the axial piston pump has been removed ad disassembled to view and analyze the damaged zone Fig. 7 shows the observable starting wear of the slipper pad. Although the testing time is short compared to the normal working time of the pump in companies, the results of this simulation can quickly lead to understanding that the slipper and the retainer are exposed to failure. One reason is a malfunction caused by the lack of sufficient spacing between the components (slipper/retainer), including the problem related to the control not being neglected in this mechanism of failure and rapid damage in hydraulic machines.

By zooming the slipper pad, we can easily see how the fatigue has already begun., it is already presenting small cracks (see Fig. 7).

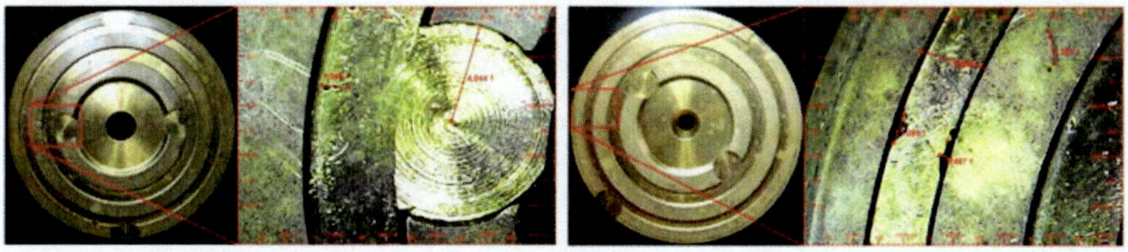

Fig. 7 The surface of the slipper pad after the test

4 Conclusion

The fracture and damage mechanism of the slipper/retainer in the axial piston machine have been elaborated in this work. The study of the degradation mechanism and wear of the slipper/retainer assembly is done. The reasons for wear, damage, and fatigue of the slippers and retainer causing huge loss of power or even total destruction of the pump are given. The most affected areas exposed to elastic or plastic deformation and related fatigue are determined thanks to the simulation, combined with the test carried out in the laboratory. It has been demonstrated that the deformation, fracture, or failure of the slipper and retainer come on one hand from the design issues and the material properties on the other hand. The space reserved between the slipper and the retainer holes is playing great roles. Materials with a large tensile strength value must be considered for manufacturing slipper-retainer because such materials undergo the least elastic deformation, consequently the least wear and failure.

1. References

1. Haidak G, Wang D, Shiju E, Li F (2019) The impact of the deformation phenomenon on the process of lubricating and improving the efficiency between the slipper and swashplate in axial piston machines. IEEE Access. 7:69393–69409 (2019). https://doi.org/10.1109/ACCESS.2019.2919493
2. Haidak G, Wang D, Shiju E, Liu J (2018) Study of the influence of slipper parameters on the power efficiency of axial piston pumps. Adv Mech Eng 10:168781401880146. https://doi.org/10.1177/1687814018801460
3. Ivantysynova M, Baker J (2009) Power loss in the lubricating gap between cylinder block and valve plate of swash plate type axial piston machines. Int J Fluid Power 10:29–43. https://doi.org/10.1080/14399776.2009.10780976
4. Schenk A, Ivantysynova M (2015) A transient thermoelastohydrodynamic lubrication model for the slipper/swashplate in axial piston machines. J Tribol 137:031701. https://doi.org/10.1115/1.4029674
5. Wieczorek U, Ivantysynova M (2002) Computer aided optimization of bearing and sealing gaps in hydrostatic machines—the simulation tool caspar. Int J Fluid Power 3:7–20. https://doi.org/10.1080/14399776.2002.10781124
6. Lengiewicz J, Stupkiewicz S (2013) Efficient model of evolution of wear in quasi-steady-state sliding contacts. Wear 303:611–621. https://doi.org/10.1016/j.wear.2013.03.051

7. Schuhler G, Jourani A, Bouvier S, Perrochat J-M (2018) Wear mechanisms in contacts involving slippers in axial piston pumps: a multi-technical analysis. J Mater Eng Perform 27:5395–5405. https://doi.org/10.1007/s11665-018-3610-5
8. Schuhler G, Jourani A, Bouvier S, Perrochat J-M (2018) Efficacy of coatings and thermo-chemical treatments to improve wear resistance of axial piston pumps. Tribol Int 126:376–385. https://doi.org/10.1016/j.triboint.2018.05.007
9. Rybiak R, Fouvry S, Bonnet B (2010) Fretting wear of stainless steels under variable temperature conditions: Introduction of a 'composite' wear law. Wear 268:413–423. https://doi.org/10.1016/j.wear.2009.08.029
10. Meneghetti G, Rigon D, Gennari C (2019) An analysis of defects influence on axial fatigue strength of maraging steel specimens produced by additive manufacturing. Int J Fatigue 118:54–64. https://doi.org/10.1016/j.ijfatigue.2018.08.034
11. Haidak G, Wang D, Shiju E (2019) Research on the thermo-elastic deformation and fracture mechanism of the slipper retainer in the axial piston pumps and motors. Eng Fail Anal. https://doi.org/10.1016/j.engfailanal.2019.02.041
12. Haidak G, Wang D, Ekemeyong Awong EL (2020) Modelling of deformation and failure of slipper-retainer assembly in axial piston machine. Eng Fail Anal 111:104490. https://doi.org/10.1016/j.engfailanal.2020.104490

Effect of Combined Substance Sorption on the Mechanical Performance of Carbon Fibre-Reinforced Polymer Composites

A. Mostafa, M. J. Lavelle, S. A. Hadigheh, K. Shankar, Y. Y. Lim, and N. Sirach

Abstract Carbon fibre-reinforced polymer (CFRP) composites are used extensively within the marine, oil and gas, aerospace, and automotive industries. In these applications, CFRP components are exposed to a wide range of corrosive substances, such as moisture, saltwater/seawater and industrial oil. As a result, CFRP components inevitably absorb undesirable substances throughout their service life, which subsequently leads to detrimental effects on their mechanical properties. The effects of individual interactions, as well as the sorption of various corrosive substances, is well documented in the literature. However, the combined effects of corrosive media interactions and induced damage to the mechanical performance of CFRP composites is not well known. Within this investigation, the effects of combined substance sorption on the mechanical properties of 2 × 2 twill weave carbon fibre-reinforced epoxy, polyester, and vinyl ester composites were compared against the behaviour of individual substance constituents. Three common solutions that CFRP may be exposed to in service are utilised in the current investigation, namely, water, salt water, and industrial hydraulic oil. Gravimetric, tension and flexural tests were undertaken to understand the physical effect of the sorption phenomenon and to quantify the amount of deterioration in the mechanical performance of CFRP under hot and wet conditions. It was found that a combination of water and hydraulic oil, or salt water and hydraulic oil, degraded the CFRP specimens more than any individual solution. Moreover, it was observed that epoxy-based composites were more susceptible to deterioration than polyester- and vinyl ester-based composites. These results are important in developing design guidance for CFRP exposure in different media.

A. Mostafa (✉) · M. J. Lavelle · Y. Y. Lim · N. Sirach
Faculty of Science and Engineering, Southern Cross University, Lismore, NSW 2480, Australia
e-mail: ahmed.thabet@scu.edu.au

S. A. Hadigheh
School of Civil Engineering, The University of Sydney, Lismore, NSW 2006, Australia

K. Shankar
School of Engineering and Information Technology, UNSW-Canberra, Canberra ACT 2600, Australia

M. Abdel Wahab (ed.), *Proceedings of the 2nd International Conference on Structural Damage Modelling and Assessment*, Lecture Notes in Civil Engineering 204,
https://doi.org/10.1007/978-981-16-7216-3_11

Keywords Carbon Fibre-Reinforced Polymer (CFRP) · Tension · Flexural properties · Sorption · Diffusion · Ageing · Combined exposure

1 Introduction

Fibre-reinforced polymer (FRP) composites are used widely around the world in the automotive, civil, marine, aerospace, and oil and gas industries [1–4]. These applications include spacecraft, water bearing vessels, commercial and military aircraft, as well as industrial vehicles and machinery [1, 5, 6]. In these applications, carbon fibre-reinforced polymer (CFRP) components can be exposed to a variety of substances that exist in a range of various environmental conditions. Common examples include interactions with water, sea water, hydraulic oils, various lubricants, and fuels [7–11]. Moreover, the conditions of moisture and temperature exposure are often cyclic and can cover wide moisture and temperature ranges. Through exposure to these conditions, the mechanical performance of CFRP components may be permanently degraded [12–15].

In real-world applications, CFRP composites are regularly exposed to wet environments of varying temperatures. Dewimille and Bunsell [16] investigated the accelerated ageing of glass fibre-reinforced polymer (GFRP) composites in a water solution. The induced damage due to water uptake was reversible and significantly relies on temperature. Increasing the water temperature to boiling point did not increase the rate of water absorption through the same mechanisms of lower temperature absorption. Instead, it resulted in damage mechanisms that allowed higher levels of water uptake to occur. Haskins and Kerr [17] investigated the effect of thermal aging on carbon/epoxy and carbon/polyamide composites. It was found that at temperatures of 120°C, no discernible degradation in the tensile strength was observed for 10,000 h. However, slight degradation was observed between 10,000 to 25,000 h, while significant degradation emerged after 50,000 h. This shows that CFRP composites can survive in service for extended periods of time under elevated temperatures. Moreover, this study revealed that the decline in the mechanical performance of the composites was due to oxidation of the matrix, leading to an increase in porosity and causing microcracks within the fibre/matrix interlaminar region that allows substantial humidity uptake.

Adams and Singh [18] studied the dynamic properties of carbon/epoxy laminates exposed to hot and wet conditions of steam. Carbon fibre-reinforced epoxy (CF-epoxy) composites were found to follow Fickian behaviour in their moisture absorption, where the water molecules were randomly infiltrated into the surface of the composite laminates, causing reduction in the shear modulus and increasing the shear loss factor. However, once the samples were dried out, their mechanical properties returned to the preconditioned levels. Conversely, carbon/epoxy composites experienced significant permanent damage when exposed to high-temperature steam, causing severe loss of fibre-matrix interfacial adhesion due to matrix plasticisation. The experimental investigation undertaken by Selzer and Friedrich [12] explored the effects of moisture absorption on the mechanical properties of CFRP composites. They reported that exposure to moisture affected the matrix-based properties and caused a reduction in fibre-matrix interfacial strength. A change in mechanical performance was only observed for those samples orientated perpendicular to the load direction. Water temperature had no influence on the extent of mechanical degradation below the glass transition temperature of the CFRP. Moreover, once the samples were fully saturated with water, the mechanical properties of the samples did not change with time over three years of exposure.

CFRP composites are regularly employed in marine applications and, as a result, are regularly exposed to salt water conditions. Liao et al. [19] conducted an investigation on the change in properties of pultruded GFRP plates after undergoing environmental aging. The flexural and tensile properties were found to degrade when exposed to either water or salt water. The salt water concentration did not appear to affect the flexural performance of the samples. In addition, the study highlighted an important comparison between accelerated aging times and the normal aging processes, where the degradation of samples exposed to 750°C water for 2,400 h was observed to be the same as samples exposed to room temperature water for 3,900 h. Kootsookos and Mouritz's [20] study was one of the first investigations on the difference in durability of glass and carbon fibre-reinforced polyester and vinyl ester composites under seawater exposure. This study concluded that polyester resin composites are less stable in salt water than vinyl ester resin composites in terms of chemistry and susceptibility to hydrolysis. The flexural properties of both the polyester and vinyl ester resins exhibited the same level of degradation with their mode I interlaminar fracture toughness manifested minor change. Moreover, the decrease in the flexural stiffness and strength was found to be in proportion to the mass uptake of salt water.

The effect of water uptake on the interlaminar and flexural performance of carbon fibre-reinforced epoxy and non-epoxy resins was investigated by Dell'Anno and Lees [21]. The epoxy systems investigated tended to be more susceptible to moisture exposure than non-epoxy resins such as vinyl esters and urethane acrylates. Moreover, proportionality was observed between the mass of water absorbed and the strength loss due to moisture effects on the resin matrix and the fibre/matrix adhesion. The epoxy-based composites showed proportionally greater strength characteristics in their dry state, suggesting that for service in moist environments, non-epoxy resins or a protective coating should be considered. Differences between fresh water and salt water absorption were also witnessed by Zafar et al. [22] when investigating the long-term exposure on carbon fibre/epoxy composite and pure epoxy to better distinguish between water and salt water effects on both the composite and its matrix. Samples exposed to salt water experienced a lower mass uptake than those exposed to fresh water. However, salt water appears to have more detrimental effects on the interfacial bonding between the matrix and the fibres than fresh water, and this effect was seen to be greater for longer exposure periods [17, 23]. Results indicated that the degradation was caused by a decrease in crosslinks within the composite due to bonding between the corrosive media molecules and the composite and epoxy matrix. This led to an inevitable reduction in the overall rigidity.

One of the few direct comparisons between the degrading effects of water and hydraulic oil on FRP composites was performed by Sala [7], where the effects of water, hydraulic oil, jet fuel, and dichloromethane on aramid/epoxy and carbon/epoxy composites were investigated. It has been found that jet fuel and de-icing fluid only affect the compressive performance, whereas the hydraulic oil showed similar effects to water exposure but with greater severity. Additionally, increasing the exposed temperature of the hydraulic oil specimens further decreased their mechanical performance. The investigation undertaken by Sugita et al. [8] explored the effects on mechanical performance of environmentally and chemically degraded CFRP lap joints commonly used within the aerospace industry. The substances used for sample immersion were water, anti-icing fluid, jet fuel, and hydraulic oil. Accelerated aging techniques were used on the test samples, causing observable chemical and irreversible hardness degradation to the anti-icing fluid and hydraulic oil specimens. The experimental results indicated that the composites anti-icing fluid and hydraulic oil uptake was significantly greater than the water and jet fuel uptake.

Mahieux et al. [24] provided insightful investigation into the effects of industrial oil and water on FRP composites in terms of absorption rate and viscoelastic behaviour. CFRP samples with different polymer matrices were immersed in either water or VG 46 industrial oil solutions at varying temperatures. The majority of samples submerged in water showed minimal water uptake, whereas all samples immersed in the oil saw no significant change in mass. This is due to the large oil molecules being unable to diffuse into the fibre/matrix structure. Moreover, dynamic mechanical analysis (DMA) was performed on the samples, and it was seen after several months of oil and water immersion that most samples had no changes to their results. Hadigheh and Ke [25, 26] proposed one-dimensional (1D) and three-dimensional (3D) diffusion models for FRP specimens exposed to aggressive acid agents. Analysis of the deterioration mechanisms of FRP specimens revealed that the elastic modulus of FRP composites slightly improved in the early stages of exposure due to the occurrence of additional cross-linking in the thermosetting resin before full curing. These additional cross-links can decrease the chain mobility and improve the mechanical properties of the composite. However, the mechanical properties of the composites decreased after this threshold time. Hadigheh and Ke's models [25, 26] showed that acid diffusion decreased with penetration depth and exposure time. The diffusion was found to be restrained within the direction perpendicular to the FRP plate, while acid penetration occurred faster along the FRP length and width.

In general, similar mechanical performance effects were observed for CFRP composites exposed to water and salt water, whereas exposure to hydraulic oil produces notable differences [22, 27–29]. Exposing CFRP to water-based liquids causes the composite matrix to swell and plasticise, thereby reducing the interfacial adhesion between the matrix and fibre. This effectively deteriorates the mechanical performance of the composite [30, 31]. By comparison, exposure to hydraulic oil is shown to lower the glass-transition temperature of the composite, resulting in greater degradation in mechanical performance at elevated temperatures [24]. In addition, the presence of water may interact with the composite in such a way that allows the absorption of larger oil molecules into the composite. The presence of a combination of these two substances can be seen within a range of applications of CFRP composites, such as in unmanned aerial vehicles (UAVs). Performing further investigations into these different interactions will enable more informed design and engineering of CFRP components within their intended uses and applications.

Although detailed investigations have been performed on individual substance absorption by CFRP composites, investigations on combined sorption affects cannot be readily found in the literature. Therefore, the main objective of this investigation is to quantify the degradation of the mechanical behaviour of CFRP laminates due to sorption of individual and combined corrosive substances that are commonly experienced in service. Since the list of possible corrosive environments in service is extensive, some of the most commonly found substances were chosen for this study, namely water, salt water, and hydraulic oil. To achieve the objectives of the study, tensile and flexural tests on CFRP laminates with different resin types were undertaken. The CFRP laminates were subjected to the following corrosive environments: demineralised water, sea water, hydraulic oil, combined water and hydraulic oil, and combined salt water and hydraulic oil.

2 Experimental Methods

2.1 Materials and Sample Preparation

In this study, all samples were prepared using 2 × 2 twill weave carbon fibre supplied by CG Composites. Three polymer resins were used to create three separate CFRP panels for testing, namely (i) Sicomin SR8100 epoxy resin and Sicomin SD 8824 hardener from Lavender Composites (resin-to-hardener ratio of 100:22), (ii) CDR 251-60GP polyester resin from NCS Composites and M.E.K.P catalyst from CG Composites (with 2% catalyst), and (iii) 700–50 vinyl ester resin from NCS Composites and M.E.K.P catalyst from CG Composites (with 2% catalyst).

The panels were prepared using the wet layup process by impregnating resin into six layers of 700 mm × 650 mm carbon fibre sheets. Resin was carefully squeezed through each layer prior to adding an additional layer and additional resin on top. Care was taken to ensure that air bubbles were pressed out, each fabric ply achieved full resin saturation, and each layer was aligned. Once the resin was impregnated through all six layers, the vacuum bagging technique was used to cure the panels. A BriskHeat Model VT4000 numerical control vacuum debulking table was used to apply a uniform pressure to the panel. To achieve this, peel ply, release fabric, and breather fabric were used in the vacuum bagging process. The wet layup and vacuum bagging processes are shown in Fig. 1. After the panels were left to cure under vacuum for 24 h, the panels were removed from the table and were further left to cure for at least three days in the laboratory. Tension and flexural specimens were cut from the cured CFRP panels using a diamond blade drop-saw.

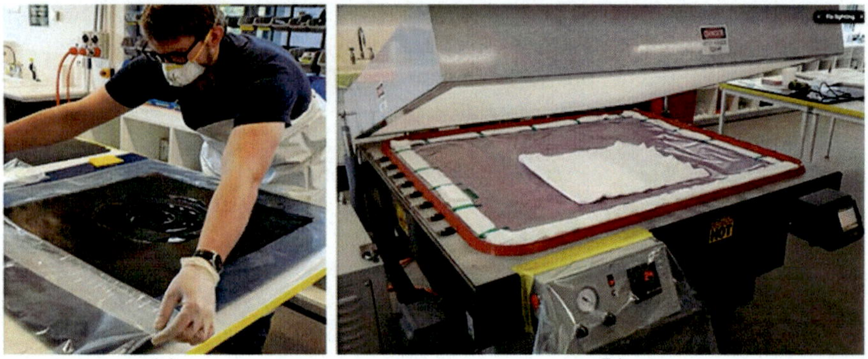

(a) (b)

Fig. 1 Manufacturing of CFRP panels: **a** resin impregnation; **b** vacuum bagging using the debulking table

A total of four individual substances were utilised in combination to produce the test solutions used in this study. These substances included demineralised water, salt water, Penrite Indus Pro ISO VG46 Hydraulic Oil, and Polysorbate 85 emulsifier. The salinity of the salt water, which was collected from a river, was measured to be approximately 33 ppt using a calibrated EUTECH Instruments Conductivity Probe with 7.83 unit pH [32]. The solutions were prepared and stored in aluminium trays, and the CFRP specimens were then placed into these trays for conditioning. The compositions of the solutions are shown in Table 1.

Accelerated aging was employed in order to simulate long-term exposure to the solutions. Test groups 1–6 were placed in to a Thermolite dehydrating oven (see Fig. 2a) at 600 °C for 25 days, whereas test group 7 was left in laboratory conditions. Thereafter, all groups were taken from the oven for mass analysis over a period of 24 h. Since several specimens were exposed to oil during the conditioning process, all specimens were subject to an acetone bath to clean the specimens and maintain consistency across the different substance exposures. After the acetone had evaporated from the surface of the specimen (one hour after exposure), the mass of the

Table 1 Corrosive substance concentration for CFRP specimen

Substance	Substance concentration (ml)						
	Group 1	Group 2	Group 3	Group 4	Group 5	Group 6[1]	Group 7[2]
Demineralised water	700	0	0	350	0	0	0
Salt water	0	700	0	0	350	0	0
Hydraulic oil	0	0	700	350	350	0	0
Emulsifier	0	0	0	80	50	0	0

[1]Aged control sample (CFRP sample subjected to only thermal aging). [2]Control sample (pristine CFRP sample)

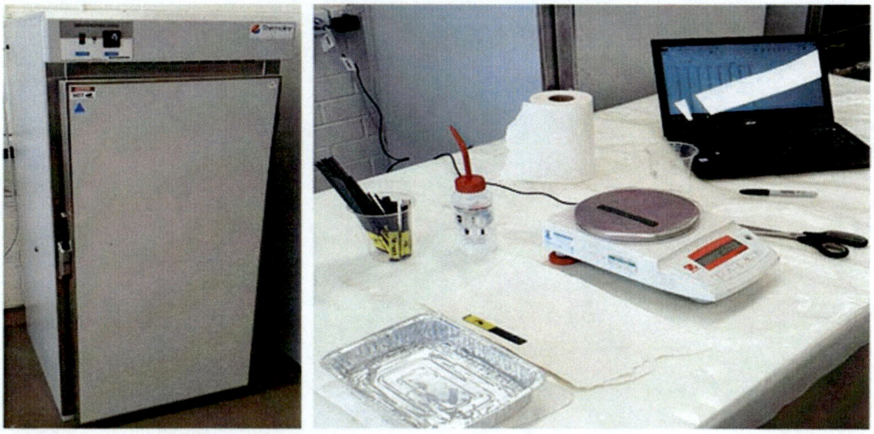

(a) (b)

Fig. 2 Preparation of CFRP specimens: **a** dehydrating oven; **b** gravimetric analysis setup

individually marked sample was recorded with a high precision scale, as illustrated in Fig. 2b.

2.2 Mechanical Testing

Tension Testing

Tension testing was performed in accordance with ASTM-D3039M [33] using a 50 kN MTS Criterion electromechanical universal testing machine (UTM). The tests were operated under a displacement-control setting with a constant crosshead speed of 2 mm/min. The dimensions of the tensile coupons were 200 mm (long) × 20 mm (wide) × 1.5 mm (thick). Aluminium tabs with a length of 40 mm and width of 20 mm were bonded onto the ends of each coupon using a two-part epoxy adhesive to prevent any local damage when clamped in the testing machine. One unidirectional strain gauge with a gauge length of 5 mm and a gauge resistance of 120 Ω was positioned at the mid-length of the specimen and was used to measure the strain on the surface of the tension coupon. All loads, displacements and strains were recoded using the built-in data acquisition (DAQ) system in the MTS UTM. Figure 3 shows the tensile coupon preparation and the test setup.

Flexural Testing

Flexural testing was carried out in accordance with ASTM-D7264M [34] using a 1 kN MTS Criterion UTM. The tests were operated under a displacement-control setting with a constant crosshead speed of 1 mm/min. The flexural coupons had

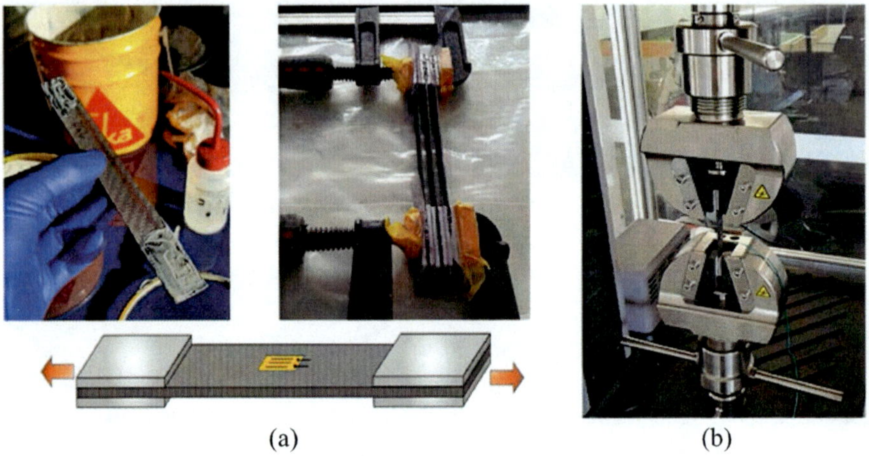

Fig. 3 CFRP tensile testing a specimen preparation, and b test setup

nominal dimensions of 108 mm (long) × 13 mm (wide) × 1.5 mm (thick). A three-point bending scheme was used in this study, with a 45 mm shear span and a centrally applied load (see Fig. 4). All loads and displacements were simultaneously recorded by a data logging system. Care was taken when adjusting the loading roller before testing to ensure that it was flush with the top surface of the specimen to accurately capture the true deflection at midspan.

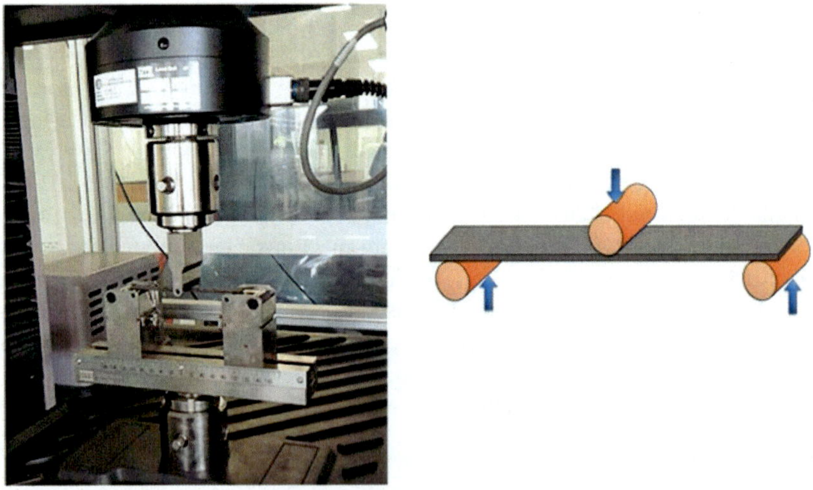

Fig. 4 Flexural testing setup

3 Results and Discussion

3.1 Gravimetric Analysis

The corrosive substance content within the investigated CFRP composites was determined using the gravimetric method. Figures 5, 6, and 7 show the average percentage of corrosive media content for each conditioned set of specimens compared to the control specimens (i.e. without ageing or exposure to corrosive medium). By comparing these figures, it is evident that the mass gain of the corrosive media varies across the three investigated resin systems used in CFRP laminates (i.e. epoxy, polyester and viny ester) due to the fluctuating physical absorption of the substances molecules into the resin system. Moreover, for epoxy, polyester and vinyl ester cured CFRP composites, the specimens exposed to water and salt water absorbed a similar

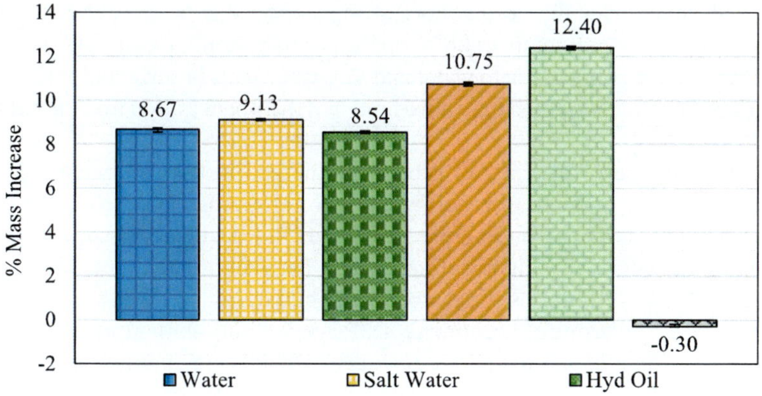

Fig. 5 Percentage mass gain for CF-epoxy specimens

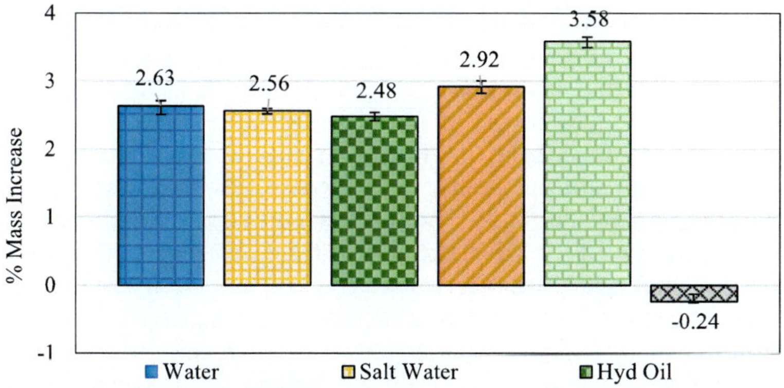

Fig. 6 Percentage mass gain for CF-Polyester specimens

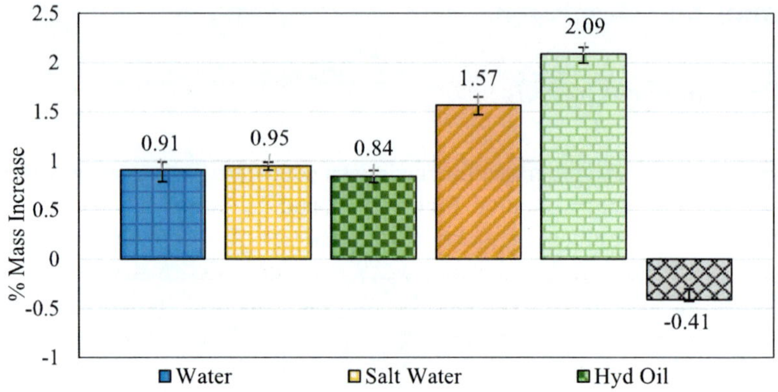

Fig. 7 Percentage mass gain for CF-vinyl ester specimens

mass of substance over the accelerated ageing period due to the close similarity between the molecular structure of both substances. In general, the samples absorbed less hydraulic oil in comparison to water and salt water. This is attributed to the larger size of the oil molecules, which may have hindered their penetration into the polymer chain.

Regardless of the resin type, all groups submerged in the combined solutions (i.e. water/hydraulic oil and salt water/hydraulic oil) absorbed larger amounts of the corrosive media in comparison to those submerged in the individual solutions. This is caused by the aggressive chemical interaction between the hydraulic oil and the resin matrix, which plasticises the matrix and facilitates substance ingress. On the other hand, the aged control specimens lost a very small amount of mass due to evaporation of some of the matrix's constituents. It is worth noting that the vinyl ester-cured CFRP specimens absorbed the least amount of corrosive media across all test groups, whereas the epoxy-cured CFRP specimens absorbed the most.

Typically, when a CFRP laminate is exposed to a corrosive environment, the majority of the corrosive media uptake is absorbed by the matrix system and only a minor amount is absorbed by the fibre [35]. The substance sorption mechanisms can be categorised as either bound or free. The bound mechanism, or hydrolysis (i.e. plasticisation), is due to the chemical bonding between the hydroxide ions in the substance and the base matrix of the CFRP, which leads to matrix degradation [36, 37]. On the other hand, the free mechanism does not require chemical interaction as the substances molecules freely move within the voids, or interfacial boundaries, of the matrix [36, 38, 39]. Two pathways are utilised to introduce the free substance ingress into CFRP laminates, namely the diffusion action (modelled by Fick's Law), or preferable paths (i.e. cavities, crevices, interfacial delamination and capillary process) due to CFRP's fabrication and service imperfections. In general, corrosive media can enter a CFRP laminate by one, or a combination of, the following mechanisms: (i) attacking the resin matrix; (ii) attacking the fibre/matrix interface;

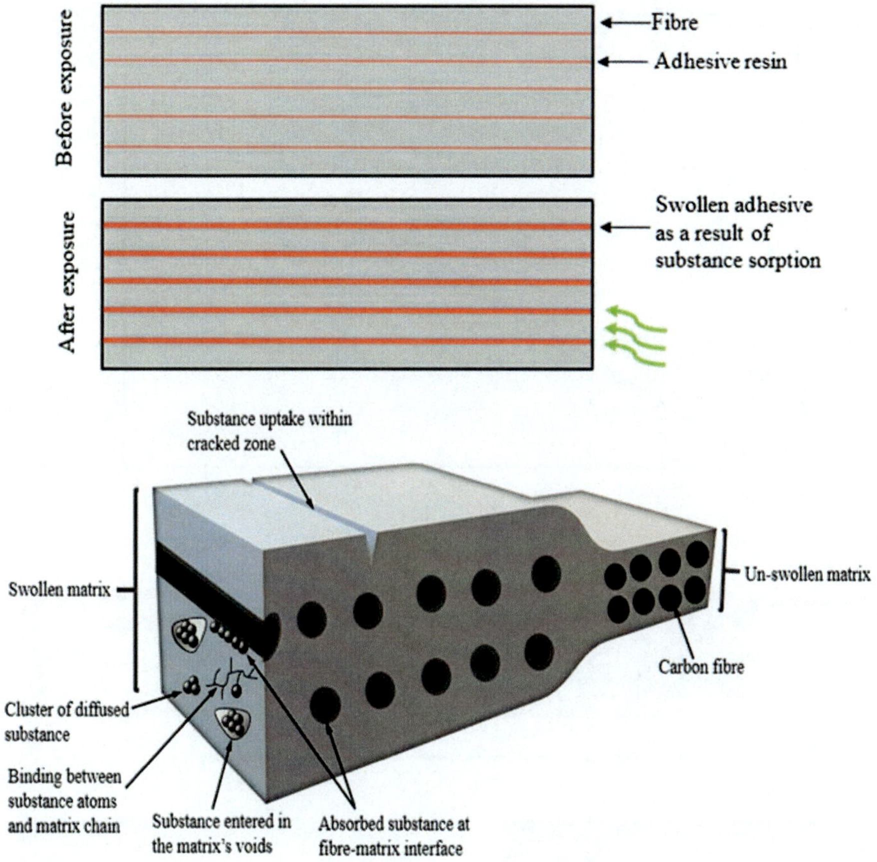

Fig. 8 Sorption mechanisms of corrosive media into CFRP specimen

and (iii) attacking the fibre [40]. Different sorption sites and mechanisms for the transportation of corrosive media into CFRP composites are illustrated in Fig. 8.

3.2 *Tensile Testing*

The tensile stress versus strain response for the different CFRP composites subjected to different corrosive media is presented in Figs. 9, 10, 11, 12 Key results from these tensile tests are presented in Tables 2, 3, and 4, which includes the elastic modulus, the ultimate tensile strength, and the ultimate strain. For each specimen variable (i.e. CF-Epoxy / Water), the average value calculated from the five test specimens is provided.

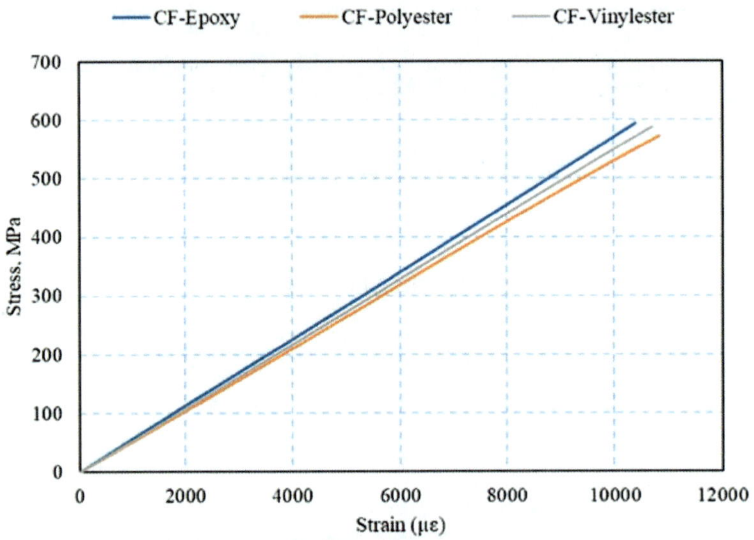

(a)

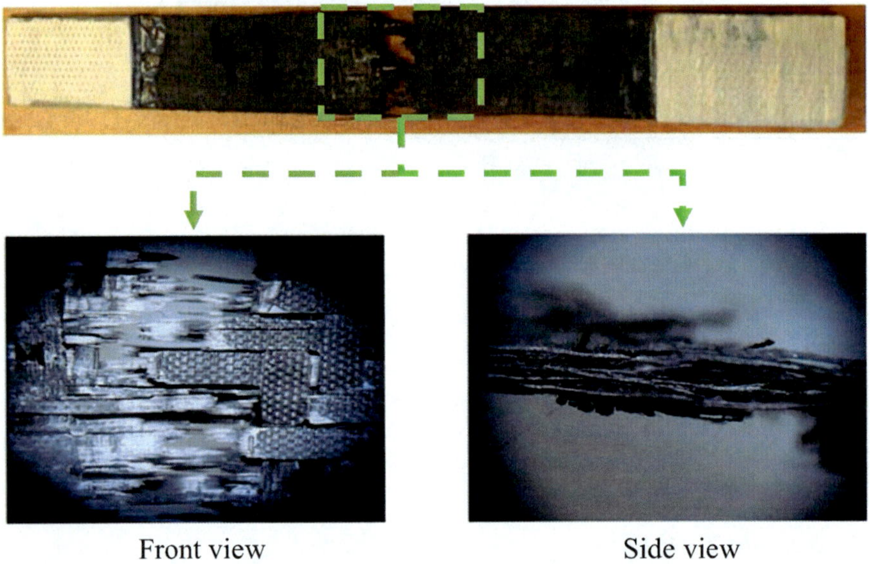

(b)

Fig. 9 **a** Tensile stress–strain relationship of CFRP composites with various resin matrices, and **b** tensile damage mode

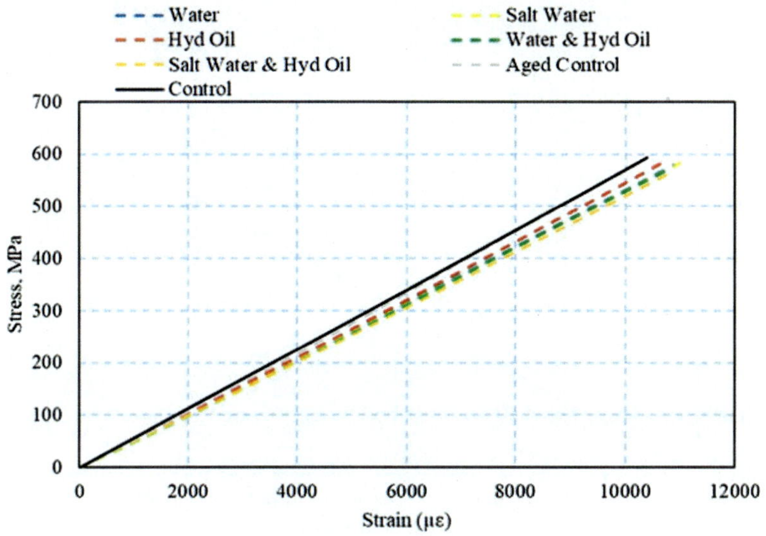

Fig. 10 Tensile stress–strain relationship of CF-epoxy composites exposed to various corrosive media

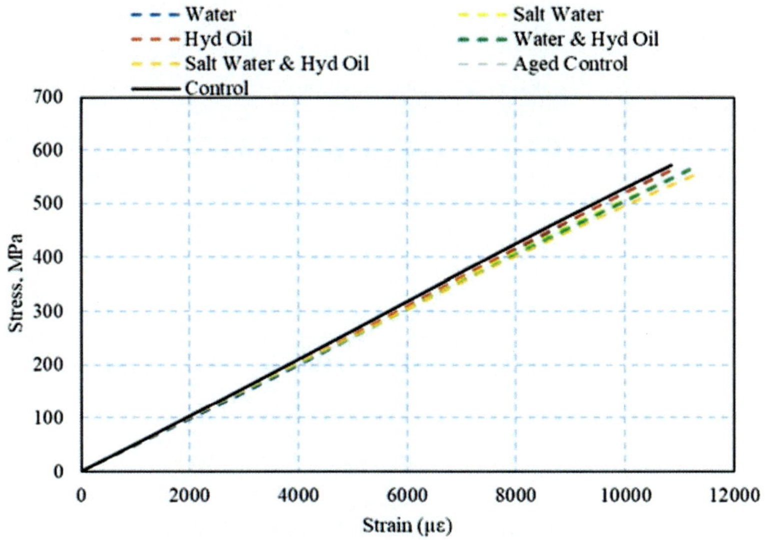

Fig. 11 Tensile stress–strain relationship of CF-polyester composites exposed to various corrosive media

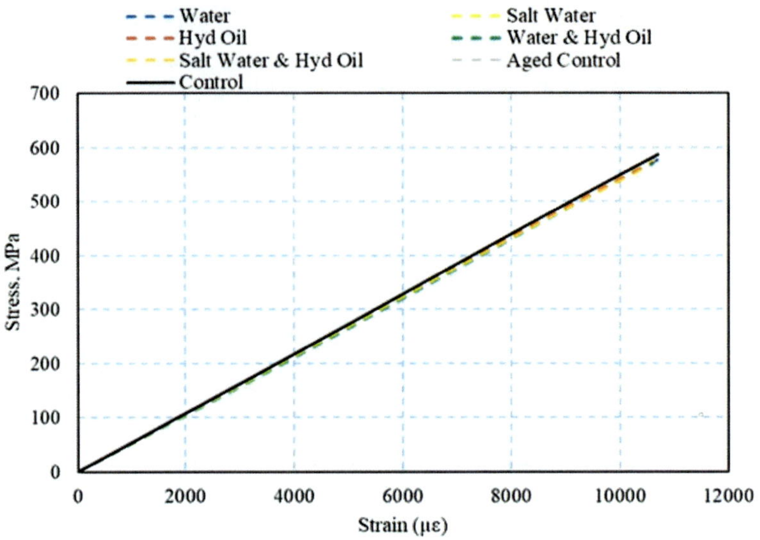

Fig. 12 Tensile stress–strain relationship of CF-vinyl ester composites exposed to various corrosive media

Table 2 Corrosive substance concentrations for CFRP specimens manufactured with CF-epoxy

	Elastic modulus		Ultimate strength		Ultimate strain	
	GPa	%[1]	MPa	%[1]	με	%[1]
Water	52.9	6.87	580.4	2.13	10,900	4.92
Salt water	52.7	7.22	583.6	1.60	11,000	5.88
Hydraulic oil	54.0	4.93	587.7	0.90	10,753	3.51
Water & Hydraulic oil	52.3	7.92	568.4	4.16	10,763	3.60
Salt water & Hydraulic oil	51.6	9.15	564.8	4.76	10,815	4.10
Aged control	56.3	0.88	591.3	0.29	10,347	0.40
Control	56.8	–	593.1	–	10,389	–

[1]% Reduction compared to control specimen

A comparison of the tensile stress–strain response of CFRP control samples with different resin matrices is presented in Fig. 9a. Due to the similarity in the responses for each CFRP-resin type, it is evident that the stiffness and strength of CFRP composites under longitudinal tension are dominated by the properties of the carbon fibres. The carbon fibres failed since they have a lower ultimate strain in comparison to the resin matrix (Fig. 9b). From Fig. 9 (a), it is evident that the epoxy-based composites have the highest elastic modulus and tensile strength, followed by the vinyl ester-based composites, and then the polyester-based composites.

The influence that exposure to different corrosive mediums has on the tensile stress–strain response of CF-epoxy laminates is shown in Fig. 10. The combined

Table 3 Corrosive substance concentrations for CFRP specimens manufactured with CF-epoxy

	Elastic modulus		Ultimate strength		Ultimate strain	
	GPa	%[1]	MPa	%[1]	με	%[1]
Water	51.6	2.46	558.6	2.28	10,638	1.90
Salt water	51.1	3.40	543.1	5.00	10,777	0.61
Hydraulic oil	51.9	1.89	560.2	2.01	10,815	0.26
Water & Hydraulic oil	50.4	4.73	562.9	1.54	11,182	3.13
Salt water & Hydraulic oil	49.9	5.67	560.5	1.95	11,433	5.44
Aged control	52.5	0.76	567.3	0.76	10,814	0.27
Control	52.9	–	571.7	–	10,843	–

[1]% Reduction compared to control specimen

Table 4 Corrosive substance concentrations for CFRP specimens manufactured with CF-epoxy

	Elastic modulus		Ultimate strength		Ultimate strain	
	GPa	%[1]	MPa	%[1]	με	%[1]
Water	53.9	1.64	580.8	1.02	10,750	0.47
Salt water	54.0	1.46	574.9	2.02	10,608	0.86
Hydraulic oil	53.8	1.82	571.4	2.61	10,602	0.91
Water & Hydraulic oil	53.6	2.19	571.0	2.69	10,600	0.93
Salt water & Hydraulic oil	53.7	2.01	566.7	3.42	10,550	1.40
Aged control	54.6	0.36	584.8	0.33	10,712	0.11
Control	54.8	–	586.8	–	10,700	–

[1]% Reduction compared to control specimen

solutions of water/hydraulic oil, as well as salt water/hydraulic oil, have an average reduction in the failure stress in comparison to the control specimen of 4.16% and 4.76%, respectively. In addition, the reduction in the elastic modulus ranged between 7 and 9%, while the ultimate strain increased by approximately 4%. For those specimens immersed in individual solutions of water, salt water or hydraulic oil, there was still a reduction in the tensile properties in comparison to the control. However this reduction was not as significant as those specimens exposed to combinations of water/hydraulic oil, and salt water/hydraulic oil.

The tensile stress–strain curves for CF-polyester samples exposed to different corrosive mediums are illustrated in Fig. 11. The specimens from different test groups had comparable gradients, and the variation between different specimen groups was lower in comparison to the comparable responses in Fig. 10. Similar to those CF-epoxy specimens, the combined solution groups that contained water/hydraulic oil and salt water/hydraulic oil experienced greater reductions in stiffness in comparison to the specimens individually exposed to water, salt water and hydraulic oil. In addition, similar observations were made in Fig. 12 for the tensile stress–strain curves of the CF-vinyl ester specimens where the combined solutions experienced the greatest reductions in stiffness. When comparing resin types, the CF-vinyl ester specimens show the most resistance to the deterioration caused by the corrosive substances used in this investigation. The test results are shown in Tables 2, 3, and 4.

Overall, it was observed that the CF-epoxy composites exposed to corrosive media experienced a greater reduction in the elastic modulus when compared to the CF-polyester and CF- vinyl ester composites. The individual sorption of water, salt water and hydraulic oil has shown to have less influence on the tensile properties of CFRP laminates in comparison to the combined water/hydraulic oil and salt water/hydraulic oil. This is attributed to the interaction between the hydraulic oil and the CFRP laminates resin matrices, whereby the hydraulic oil plasticises the resin and facilitates more water ingress into the laminate, particularly in the epoxy specimens [21, 41–43]. Epoxy-based composites tend to show greater deterioration in tensile properties in comparison to non-epoxy composites such as polyester and vinyl ester when they are exposed to corrosive environments.

3.3 *Flexural Testing*

Three-point bending tests were conducted to understand the influence that different corrosive environments have on the flexural response of CFRP laminates. For each flexural test, the load was applied centrally and the applied load and loading roller displacement were recorded. Polynomial fitting was used to create a single trendline for each of the tested flexural samples. The representative curves shown in Fig. 13

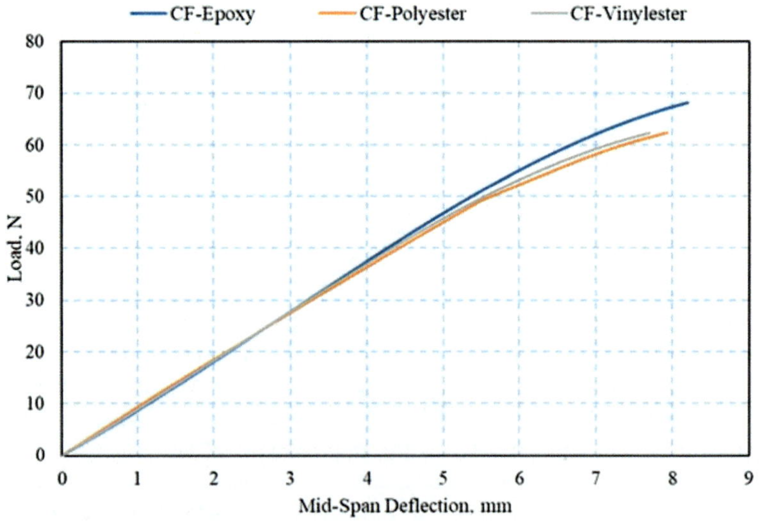

Fig. 13 Load–deflection relationship of control CFRP composites with various resin matrices

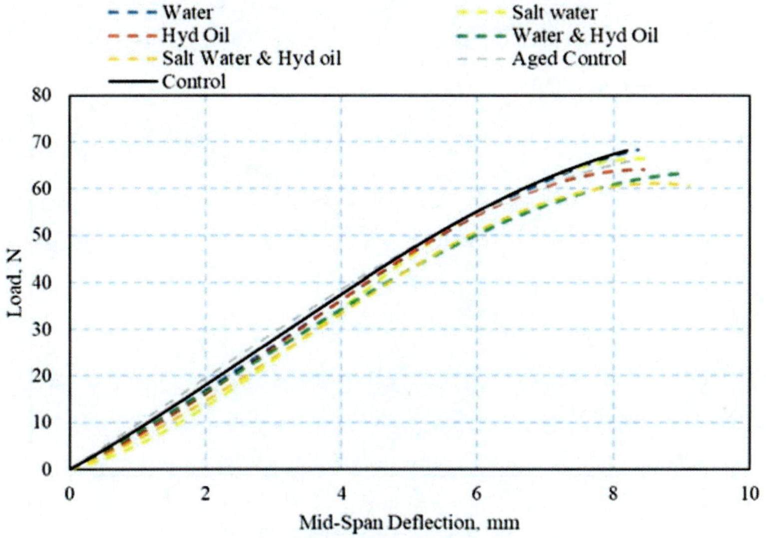

Fig. 14 Load–deflection relationship of CF-epoxy composites exposed to various corrosive media

illustrate the behaviour of control CFRP composites fabricated using epoxy, polyester and vinyl ester resin matrices. Although all curves exhibited a very similar initial slope, the ultimate load carrying capacity of the CF-epoxy composites exceeded those of the CF-vinyl ester and CF-polyester samples.

The specimens exposed to corrosive environments showed varying levels of deterioration depending on the type of resin matrix used in the laminate and the type of corrosive media they were exposed to. Figure 14 shows the load–deflection curves for the CF-epoxy flexural specimens exposed to different corrosive environments. It is shown that the samples exposed to individual solutions of water, salt water and hydraulic oil all had similar failure loads and deflections. This behaviour is expected since the sorption of these substances is quite similar. However, the samples exposed to the combination of both water/hydraulic oil and salt water/hydraulic oil exhibited lower failure loads and higher deflections than those exposed to the individual substances. A similar trend was observed for the CF-polyester specimens (see Fig. 15) and the CF-vinyl ester specimens (see Fig. 16), however the influence of corrosive environments is less pronounced on CF-polyester specimens and even less pronounced on CF-vinyl ester specimens. The reduction in strength and stiffness in the flexural response of CFRP composites is a result of the deterioration in the interlaminar shear strength of the CFR due to swelling of the resin matrix [44].

The type of matrix used in the CFRP laminate has little influence on the damage mode of the control specimens under bending. From Fig. 17a, it can be seen that the failure of the control samples is caused by fibre rupture. However, for those specimens that were exposed to corrosive environments, the damage mode changed from fibre rupture to fibre delamination. The substance sorption for all specimens

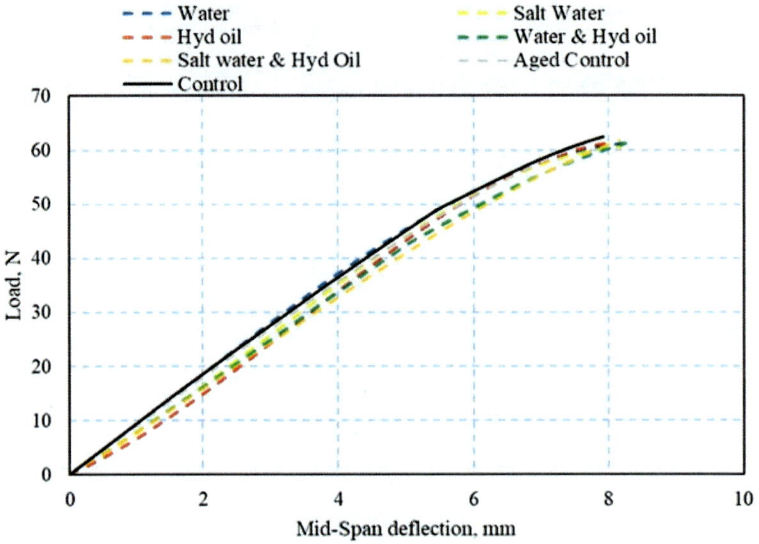

Fig. 15 Load–deflection relationship of CF-polyester composites exposed to various corrosive media

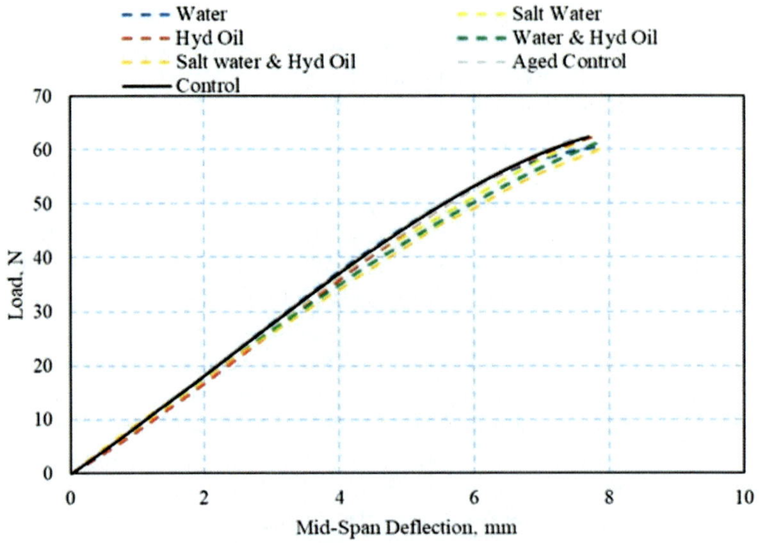

Fig. 16 Load–deflection relationship of CF-vinyl ester composites exposed to various corrosive media

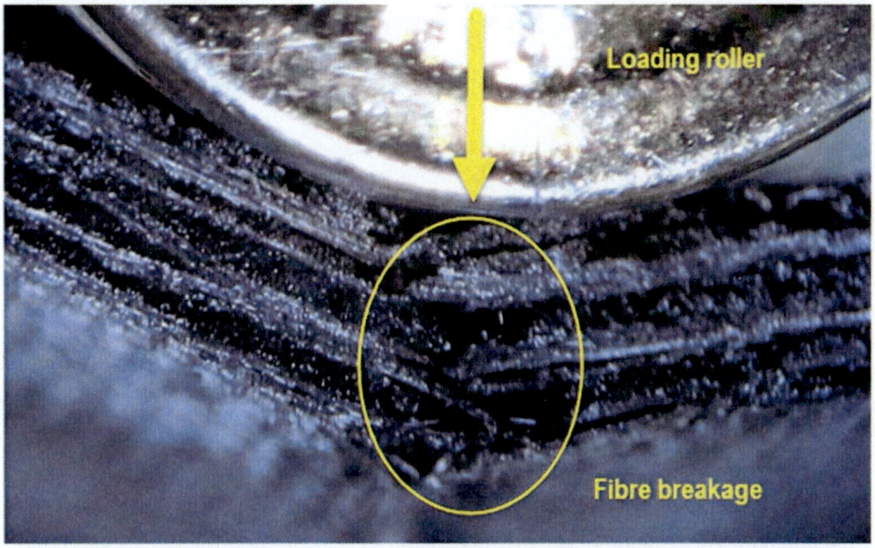

(a)

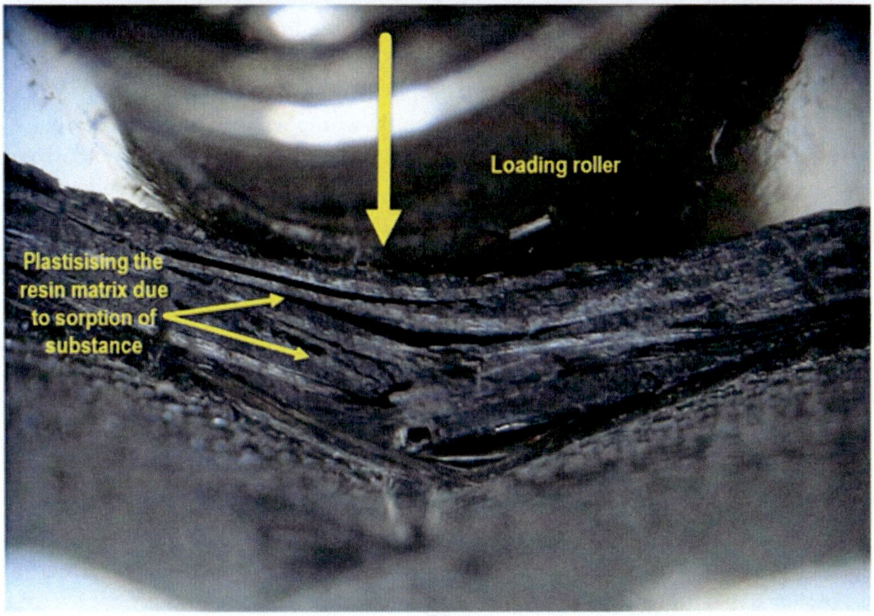

(b)

Fig. 17 Flexural damage mode of CFRP composites (**a**) control specimen, and (**b**) specimen exposed to various corrosive media exposed to corrosive media altered the damage mode as a result of plasticising the matrix resin (Fig. 17b).

4 Conclusions

In this study, the effects of combined substance sorption on the tensile and flexural behaviour of 2 × 2 twill weave CFR epoxy, polyester, and vinyl ester composites were compared with individual substance sorption effects. The corrosive environments included combinations of water, salt water, and hydraulic oil. Based on the results obtained from the test procedure, the following conclusions can be made:

1. The gravimetric analysis results showed that the CF-epoxy specimens absorbed more solution mass across all tested groups when compared to both the CF-polyester and CF-vinyl ester specimens. Additionally, the corrosive substance content within the CF-polyester specimens was higher than the equivalent CF-vinyl ester specimens.
2. The resin matrix used to develop the CFRP laminate has a minor influence on the tensile and flexural behaviour. The CF-epoxy composites are slightly stiffer and stronger than the CF-vinyl ester composites, which are in turn slightly stiffer and stronger than the CF-polyester composites.
3. The tensile stress–strain curves of CF-epoxy composites exposed to corrosive mediums displayed a greater reduction in mechanical properties in comparison to equivalent CF-polyester specimens. Moreover, the CF-polyester composites exhibited greater reduction in mechanical properties in comparison to the CF-vinyl ester specimens. CF-polyester composites displayed more variability in the initial slope of the stress–strain curve than CF-vinyl ester.
4. The CF-epoxy specimens exposed to corrosive environments had a greater reduction in load carrying capacity under flexure when compared to equivalent CF-polyester and CF-vinyl ester specimens. The load carrying capacity of CF-vinyl ester specimens was least affected by corrosive environments.
5. Regardless of the resin type, CFRP laminates exposed to combined solutions of both water/hydraulic oil and salt water/hydraulic oil had a greater deterioration in mechanical properties in comparison to those specimens exposed to individual solutions of water, salt water or hydraulic oil. Although specimens submerged with combined solutions exhibited a lower ultimate strength, they experienced higher deflections. In addition, the CF-epoxy specimens showed higher sensitivity to water ingress in comparison to the CF-polyester and CF-vinyl ester specimens.
6. The matrix-dominated properties of the CFRP laminates (i.e. flexural strength) were more influenced by substance absorption than the fibre-dominated properties (i.e. tensile strength).

Conclusively, CF-polyester and CF-vinyl ester composites should be considered for applications where the composites are likely to be exposed to hot and wet conditions. In the case of CF-epoxy, an adequate protective shield should be utilised during the design of the composite laminate to prevent any deterioration in the structural performance due to sorption of the surrounding media. Further research on the effect of combined substance sorption is recommended.

References

1. Selvaraju S, Ilaiyavel (2011) Applications of composites in marine industry. J Eng Res Stud 2(2):89–91
2. Mostafa A (2019) Experimental and numerical investigation on enhancing the structural integrity of composite sandwich structure. Adv Struct Eng 22(9):2149–2162
3. Sirach N, Smith ST, Yu T, Mostafa A (2021) Experimental study on the confinement of concrete cylinders with large rupture-strain FRP composites. J Compos Constr 25(4):04021026
4. Saafi M (2000) Design and fabrication of FRP grids for aerospace and civil engineering applications. J Aerosp Eng 13(4):144–149
5. Fitzer E (1985) Carbon fibres and their composites, 1st edn. Springer, Heidelberg
6. Mortensen A (2006) Concise encyclopedia of composite materials. Elsevier Science
7. Sala G (2000) Composite degradation due to fluid absorption. Compos B Eng 31(5):357–373
8. Sugita Y, Winkelmann C, La Saponara V (2010) Environmental and chemical degradation of carbon/epoxy lap joints for aerospace applications, and effects on their mechanical performance. Compos Sci Technol 70(5):829–839
9. Aktaş A, Uzun İ (2008) Sea water effect on pinned-joint glass fibre composite materials. Compos Struct 85(1):59–63
10. Jedidi J, Jacquemin F, Vautrin A (2005) Design of accelerated hygrothermal cycles on polymer matrix composites in the case of a supersonic aircraft. Compos Struct 68(4):429–437
11. Starkova O, Aniskevich K, Sevcenko J (2020) Long-term moisture absorption and durability of FRP pultruded rebars. Mater Today: Proc
12. Selzer R, Friedrich K (1997) Mechanical properties and failure behaviour of carbon fibre-reinforced polymer composites under the influence of moisture. Compos A Appl Sci Manuf 28(6):595–604
13. Chen Y, Davalos JF, Ray I, Kim HY (2007) Accelerated aging tests for evaluations of durability performance of FRP reinforcing bars for concrete structures. Compos Struct 78(1):101–111
14. La De Osa O, Alvarez VA, Fraga AN, Mammone EM, Vázquez A (2005) Loss of mechanical properties by water absorption of vinyl-ester reinforced with glass fiber. J Reinf Plast Compos 25(2):215–221
15. Hadigheh SA, Gravina RJ, Smith ST (2017) Effect of acid attack on FRP-to-concrete bonded interfaces. Constr Build Mater 152:285–303
16. Dewimille B, Bunsell AR (1983) Accelerated ageing of a glass fibre-reinforced epoxy resin in water. Composites 14(1):35–40
17. Kerr JR, Haskins JF (1984) Effects of 50,000 h of thermal aging on graphite/epoxy and graphite/polyimide composites. AIAA J 22(1):96–102
18. Adams RD, Singh MM (1996) The dynamic properties of fibre-reinforced polymers exposed to hot, wet conditions. Compos Sci Technol 56(8):977–997
19. Liao K, Schultheisz CR, Hunston DL (1999) Effects of environmental aging on the properties of pultruded GFRP. Compos B Eng 30(5):485–493
20. Kootsookos A, Mouritz AP (2004) Seawater durability of glass- and carbon-polymer composites. Compos Sci Technol 64(10):1503–1511

21. Dell'Anno G, Lees R (2012) Effect of water immersion on the interlaminar and flexural performance of low cost liquid resin infused carbon fabric composites. Compos B Eng 43(3):1368–1373
22. Zafar A, Bertocco F, Schjødt-Thomsen J, Rauhe JC (2012) Investigation of the long-term effects of moisture on carbon fibre and epoxy matrix composites. Compos Sci Technol 72(6):656–666
23. Bishop SM (1983) Effect of moisture on the notch sensitivity of carbon fibre composites. Composites 14(3):201–205
24. Mahieux CA, Lehmann D, desLigneris A (2002) Experimental determination of the effects of industrial oil immersion on polymer-based composites. Polymer Test 21(7):751–756
25. Hadigheh SA, Ke F (2019) Acid penetration kinetics in externally bonded FRP-strengthened reinforced concrete structures: mechanism and prediction. In: The 14th international symposium on Fiber-Reinforced Polymer Reinforcement of Concrete Structures (FRPRCS 14). Belfast, North Ireland
26. Hadigheh SA, Ke, F (2019) A 3D model of acid diffusion and mass uptake in Fibre-Reinforced Polymer (FRP) composites. In: The seventh Asia-Pacific conference on FRP in structures (APFIS 2019). Gold Coast, Australia
27. Kafodya I, Xian G, Li H (2015) Durability study of pultruded CFRP plates immersed in water and seawater under sustained bending: water uptake and effects on the mechanical properties. Compos B Eng 70:138–148
28. Saito H, Kimpara I (2009) Damage evolution behavior of CFRP laminates under post-impact fatigue with water absorption environment. Compos Sci Technol 69(6):847–855
29. Cabral-Fonseca S, Correia JR, Custódio J, Silva HM, Machado AM, Sousa J (2018) Durability of FRP—concrete bonded joints in structural rehabilitation: a review. Int J Adhes Adhes 83:153–167
30. Naceri A (2009) An analysis of moisture diffusion according to Fick's law and the tensile mechanical behavior of a glass-fabric-reinforced composite. Mech Compos Mater 45(3):331–336
31. Blackburn BP, Tatar J, Douglas EP, Hamilton HR (2015) Effects of hygrothermal conditioning on epoxy adhesives used in FRP composites. Constr Build Mater 96:679–689
32. Millero FJ, Feistel R, Wright DG, McDougall TJ (2008) The composition of standard seawater and the definition of the reference-composition salinity scale. Deep Sea Res Part I 55(1):50–72
33. ASTM-D3039/D3039M–17 (2017) Standard test method for tensile properties of polymer matrix composite materials. ASTM International, Philadelphia
34. ASTM-D7264/D7264M-15 (20105) Standard test method for flexural properties of polymer matrix composite materials. ASTM International, Philadelphia
35. Vodicka R (1998) Accelerated environmental testing of composite materials/Roger Vodicka. DSTO Aeronautical and Maritime Research Laboratory, Melbourne
36. Cahn RW, Lifshin E (1993) Concise encyclopedia of materials characterization, 1st edn. Pergamon Press, Oxford, New York
37. Singh Raman RK, Guo F, Al-Saadi S, Zhao XL, Jones R (2020) Understanding fibre-matrix degradation of FRP composites for advanced civil engineering applications: an overview. Corros Mater Degrad 1(1):27–41
38. Bonniau P, Bunsell AR (1981) A Comparative study of water absorption theories applied to glass epoxy composites. J Compos Mater 15(3):272–293
39. Shin KB, Kim CG, Hong CS (2003) Correlation of accelerated aging test to natural aging test on graphite-epoxy composite materials. J Reinf Plast Compos 22(9):849–861
40. Bond DA, Smith PA (2006) Modeling the transport of low-molecular-weight penetrants within polymer matrix composites. Appl Mech Rev 59(5):249–268
41. Vasiliev VV, Morozov EV (2013) Chapter 7—Environmental, special loading, and manufacturing effects. In: Vasiliev VV, Morozov EV (eds) Advanced mechanics of composite materials (3rd edn). Elsevier, Boston, pp 353–433
42. Xiao GZ, Shanahan MER (1997) Water absorption and desorption in an epoxy resin with degradation. J Polym Sci, Part B: Polym Phys 35(16):2659–2670

43. Gagani A, Fan Y, Muliana AH, Echtermeyer AT (2017) Micromechanical modeling of anisotropic water diffusion in glass fiber epoxy reinforced composites. J Compos Mater 52(17):2321–2335
44. Sciolti MS, Frigione M, Aiello MA (2010) Wet lay-up manufactured FRPs for concrete and masonry repair: Influence of water on the properties of composites and on their epoxy components. J Compos Constr 14(6):823–833

Adaptive Phase-Field Modeling of Brittle Fracture

Krešimir Jukić, Tomislav Jarak, Karlo Seleš, and Zdenko Tonković

Abstract The phase-field method for brittle fracture is a powerful and versatile method, but it suffers from high computational costs due to high mesh density requirement. This problem can be tackled by the development of adaptive remeshing procedures. In this paper, a simple and easily implementable adaptive remeshing scheme for a phase-field model of brittle fracture is presented. The mesh refinement scheme is based on the splitting of triangle elements into two new elements, resulting in conforming meshes. A simple mesh hierarchy is utilized for mesh coarsening. Remeshing criteria are based on physical quantities such as the deformation energy density and the gradient of phase-field. The phase-field model is solved using a staggered algorithm and the remeshing procedure is performed after every two iterations of the algorithm. The AT2 model is utilized, together with the spectral energy split. Generated results such as crack paths and force–displacement diagrams are compared to the results from the literature. It is shown that the utilization of adaptive remeshing leads to significant time savings without compromising the solution accuracy.

Keywords Phase-field method · Brittle fracture · Adaptive mesh refinement

K. Jukić · T. Jarak (✉) · K. Seleš · Z. Tonković
Faculty of Mechanical Engineering and Naval Architecture, University of Zagreb, I. Lučića 5, 10000 Zagreb, Croatia
e-mail: tomislav.jarak@fsb.hr

K. Jukić
e-mail: kresimir.jukic@fsb.hr

K. Seleš
e-mail: karlo.seles@fsb.hr

Z. Tonković
e-mail: zdenko.tonkovic@fsb.hr

M. Abdel Wahab (ed.), *Proceedings of the 2nd International Conference on Structural Damage Modelling and Assessment*, Lecture Notes in Civil Engineering 204,
https://doi.org/10.1007/978-981-16-7216-3_12

1 Introduction

One of the major concerns regarding structures is an unexpected failure which can lead to catastrophic consequences. Often, the mode of failure can be described as a brittle fracture. Traditionally, brittle fracture problems are solved using conventional methods based on the Linear-Elastic Fracture Mechanics (LEFM), or numerical methods such as the Extended Finite Element Method. Those methods are able to solve various problems associated with cracking and failure, but can still run into problems when more complex phenomena such as crack initiation, branching, and merging have to be properly addressed. By the development of the phase-field approach, those problems can be overcome. The phase-field approach to fracture is based on an approximative description of crack with the phase-field, while the surface of the crack is approximated with function inspired by Ambrosio-Tortorelli functional and crack growth is driven by the minimization of total energy functional. The resulting method is versatile and can be extended to various problems such as ductile fracture [1], dynamic fracture [2], fatigue crack growth [3], etc. Unfortunately, the method's biggest disadvantage is enormous computational requirements, mainly due to the requirement of high mesh density around the crack. In this work, we tackle this problem by developing a simple adaptive remeshing procedure.

Various adaptive approaches to the phase-field fracture have been reported. Adaptive refinement was firstly used on unstructured triangle meshes in [4], where two algorithms for 2D adaptivity were presented with the refinement being driven by an estimation of residual. One algorithm is based on remeshing after each increment of solver, while another one is based on remeshing after each iteration. Results have shown sensitivity to the algorithm's parameters. Anisotropic remeshing was presented in [5], where two similar algorithms were used. Anisotropic remeshing led to significant computational time savings in comparison to isotropic remeshing. Similar results were obtained in [6], where the problem was extended to thermally induced fracture, and in [7], where the brittle fracture was studied extensively. In [8] a predictor–corrector scheme for mesh adaptivity was used, while in [9] a residual type a posteriori error estimator was presented, and in both works, quadrangle meshes with hanging nodes were used. In [10] a remeshing is driven by deformation energy density. Adaptive remeshing was used in [11], where refinement was driven with recovery type indicator. Adaptivity for the problems of ductile fracture was presented in [12] and adaptivity for thermally induced fracture in [13]. Other notable work on this subject includes [14], where a simple algorithm for quadrangle and triangle splitting and recombination was proposed. In [15] adaptive remeshing with different meshes for the displacement and the phase-field fields are used, and in [16], adaptive remeshing was used within the virtual element framework.

In this work, a simple algorithm based on the splitting of triangle elements is presented. The main features of the algorithm are simplicity, easy implementation, and low computational costs. The algorithm relies on simple splitting and merging operations with mesh quality preservation, similarly as in [14] and in contrast to work as [5–7], where mesh rebuilding process is more complex. Such an approach leads to less computational effort in remeshing, and a remapping of both nodal and integration point data that is easier and less prone to error. Refinement and coarsening are based on intuitive physical quantities such as the deformation energy density, the phase-field gradient, and the stress error similarly as in [10]. While in most of the works error estimates with the mathematical background are used, in this work we show that a simple physical-based remeshing indicator is sufficient to ensure a proper mesh for crack growth. Also, we rely purely on triangle finite elements, which is in contrast to quadrangle used in some papers more versatile when it comes to both creation of initial mesh and remeshing. The algorithm is tested on various examples and it led to significant time savings while preserving the accuracy of the solution.

Work is organized in 5 sections. In Sect. 2 the phase-field model of fracture is described, and in Sect. 3 the adaptive remeshing algorithm is presented. In Sect. 4 results are presented together with discussion, followed by the conclusions given in Sect. 5.

2 Phase-Field Model of Brittle Fracture

In this section, the phase-field model of brittle fracture is described. The starting point for a description of the phase-field model of fracture is a variational approach to fracture. The following steps in the model derivation are the regularization of crack surface, the degradation of deformation energy, the energy splitting, and imposing the crack irreversibility condition, described in the following subsections.

2.1 The Variational Approach to Fracture

A variational approach to fracture has been proposed by Francfort and Marigo [17] and presents an extension of the Griffith's theory [18]. The Griffith's theory states that a crack grows if the total energy of a body declines with the crack growth, i.e., if the energy release rate is higher than the critical energy release rate. The variational approach generalizes the Griffith's approach by stating that fracture is governed by the minimization of total energy functional. The governing functional is given by:

$$\Psi = \Psi_\mathrm{e} + \Psi_f = \int_\Omega \psi \mathrm{d}V + \int_\Gamma G_c \mathrm{d}A, \tag{1}$$

where Ψ is the total energy, Ψ_e is the deformation energy, Ψ_f is the fracture energy, ψ is the deformation energy density, and G_c is the critical energy release rate. $\Omega \subset \mathbb{R}^n$ stands for an n-dimensional body, and $\Gamma \subset \mathbb{R}^{n-1}$ is the surface of the crack. With such description, phenomena such as crack initiation, merging, branching etc. can be handled explicitly, without any ad hoc criterion. Still, the exact description of the crack surface is cumbersome.

2.2 *Regularization of Crack Surface*

As the exact description of the crack is complex, an approximative one can be adopted. Bourdin [19] was the first to do so, by incorporating Ambrosio and Tortorelli regularization from image segmentation. The resulting approximative crack surface integral is:

$$\int_{\Gamma} \mathrm{d}A \approx \int_{\Omega} \gamma \mathrm{d}V = \int_{\Omega} \frac{1}{2}\left(\frac{1}{l}\phi^2 + l(\nabla\phi)^2\right)\mathrm{d}V, \tag{2}$$

where γ is called the crack surface density function, l is the length-scale and ϕ is the phase-field. The phase-field is a damage variable bounded by 0 and 1, where the value 0 corresponds to the non-damaged material and value 1 corresponds to the fully damaged material. Minimizing function (2) in 1D yields the following ordinary differential equation (ODE):

$$\frac{1}{l}\phi - l \cdot \phi'' = 0, \tag{3}$$

and solving that ODE with the boundary condition $\phi(0) = 1$ and $\phi(\pm\infty) = 0$ results in crack profile:

$$\phi = e^{-\frac{|x|}{l}} \tag{4}$$

The diffusiveness of the crack profile is governed by the length-scale. In contrast to the previously described function which results in the so-called AT2 model, currently various other functions are employed to define the phase field, and we refer to [20] for a detailed review.

2.3 *Deformation Energy Function*

As the growth of crack, i.e., the phase-field increases the total energy, it is necessary to degrade the deformation energy to allow the crack to initiate and grow. Degrading the deformation energy results in the following expression:

$$\Psi_e = \int_{\Omega} \psi \mathrm{d}V = \int_{\Omega} (g \cdot \psi_0^+ + \psi_0^-)\mathrm{d}V, \tag{5}$$

where g is the degradation function, ψ_0^+ is the tensile deformation energy density part and ψ_0^- is the compressive deformation energy density part. We adopt the most commonly used quadratic degradation function given by:

$$g = (1 - \phi)^2. \tag{6}$$

Also, here the Miehe's [21] spectral energy split is adopted, and hence the tensile and compressive deformation energy densities are defined as:

$$\psi_0^\pm = \frac{1}{2}\lambda \cdot \langle \mathrm{tr}\, \boldsymbol{\varepsilon} \rangle_\pm^2 + \mu \cdot \mathrm{tr}(\boldsymbol{\varepsilon}_\pm^2). \tag{7}$$

Herein, $\langle x \rangle_\pm = \frac{1}{2}(x \pm |x|)$ are the Macaulay brackets, λ and μ are the Lame's constants, $\boldsymbol{\varepsilon}$ is the strain tensor and $\boldsymbol{\varepsilon}_\pm$ stands for the positive and the negative parts of the strain tensor defined by:

$$\boldsymbol{\varepsilon}_\pm = \sum_{i=1}^{3} \langle \varepsilon_{\mathrm{i}}^* \rangle + \mathbf{n}_{\mathrm{i}}^* \otimes \mathbf{n}_{\mathrm{i}}^*, \tag{8}$$

where $\varepsilon_{\mathrm{i}}^*$ are the principal strains, i.e., the eigenvalues of the strain tensor and $\mathbf{n}_{\mathrm{i}}^*$ are the directions of the principal strains, i.e., the eigenvectors of strain tensor. As there are a number of degradation functions and energy decompositions, we again refer to review by Wu [20].

2.4 *Crack Irreversibility*

As the fracture energy is dissipative, then it is by the arguments of thermodynamics necessary to enforce its growth. Since the degradation function is monotonically decreasing, it can be shown that it is only necessary to enforce the phase-field irreversibility. There are various ways to enforce this irreversibility; from incorporating optimization techniques with constraints on the phase-field, to the implicit enforcement of irreversibility. In this work, the implicit enforcement with a history field by Miehe [22] is adopted. The history field is defined by:

$$H = \max_{\tau=[0,t]} \psi_0^+(\tau). \tag{9}$$

The variation of the total energy with respect to the phase-field, initially stated as:

$$\delta\Psi = \int_\Omega \delta\phi \frac{dg}{d\phi} \psi_0^+ dV + \int_\Omega \delta\phi \frac{d\gamma}{d\phi} dV \geq 0, \tag{10}$$

is modified to:

$$\delta\Psi = \int_{\Omega} \delta\phi \frac{dg}{d\phi} H \mathrm{d}V + \int_{\Omega} \delta\phi \frac{\mathrm{d}\gamma}{\mathrm{d}\phi} \mathrm{d}V = 0. \tag{11}$$

Note that the variational inequality (10) is modified to the variational equality (11) and hence theoretically, it is not the same problem. Still, results obtained by employing the history field show great resemblance to the results obtained by the optimization approach. A detailed review of irreversibility enforcement techniques can be found in [23].

2.5 Finite Element Discretization

As the phase-field fracture is described with partial differential equations, the Finite Element Method is the usual method to solve the problem. The displacement field and the phase-field within a finite element are described as:

$$\mathbf{u} = \mathbf{N}_u \mathbf{v}, \tag{12}$$

$$\phi = \mathbf{N}_\phi \boldsymbol{\phi}, \tag{13}$$

where $\mathbf{u}$ is the vector of displacements, $\mathbf{v}$ and $\boldsymbol{\phi}$ are the vectors of nodal displacements and nodal phase-field values, respectively, and $\mathbf{N}_u$ and $\mathbf{N}_\phi$ represent the matrices of displacement and phase-field shape functions, respectively. In this work, the second-order triangle element with 6 nodes is used, with quadratic distribution of the displacements and the phase field.

Adding the potential of external forces to the total energy, variating the total energy with respect to the nodal displacements and the nodal phase-field, setting those variations to be zero, and using the history field leads to the following equations:

$$\int_V \boldsymbol{\sigma}^{\mathrm{T}} \mathbf{B}_u \mathrm{d}V - \int_V \mathbf{b}^{\mathrm{T}} \mathbf{N}_u \mathrm{d}V - \int_B \mathbf{t}^{\mathrm{T}} \mathbf{N}_u \mathrm{d}A = 0, \tag{14}$$

$$\int_V \left(\left(\frac{dg}{\mathrm{d}\phi} H + \frac{G_\mathrm{c}}{l} \phi \right) \mathbf{N}_\phi + G_\mathrm{c}\, l\, \nabla^{\mathrm{T}} \phi\, \mathbf{B}_\phi \right) \mathrm{d}V = 0. \tag{15}$$

where $\mathbf{B}_u$ is a matrix that relates nodal displacement to deformations and $\mathbf{B}_\phi$ is matrix that relates nodal phase-field values to the gradient of phase-field. The residuals of (14) and (15) are defined as:

$$\mathbf{R}_u = -\int_V \boldsymbol{\sigma}^{\mathrm{T}} \mathbf{B}_u \mathrm{d}V + \int_V \mathbf{b}^{\mathrm{T}} \mathbf{N}_u \mathrm{d}V + \int_B \mathbf{t}^{\mathrm{T}} \mathbf{N}_u \mathrm{d}A, \tag{16}$$

$$\mathbf{R}_\phi = -\int_V \left(\left(\frac{dg}{\mathrm{d}\phi} H + \frac{G_\mathrm{c}}{l} \phi \right) \mathbf{N}_\phi + G_\mathrm{c}\, l\, \nabla^\mathrm{T} \phi\, \mathbf{B}_\phi \right) \mathrm{d}V \tag{17}$$

The corresponding stiffness matrices (the negative Jacobian of residual vectors) are:

$$\mathbf{K}_{uu} = \int_V \mathbf{B}_u^\mathrm{T} \frac{\partial \sigma}{\partial \boldsymbol{\varepsilon}} B_u dV, \tag{18}$$

$$\mathbf{K}_{\phi\phi} = \int_V \left(\mathbf{N}_\phi^\mathrm{T} \left(\frac{d^2 g}{\mathrm{d}\phi^2} H + \frac{G_\mathrm{c}}{l} \right) \mathbf{N}_\phi + G_\mathrm{c}\, l\, \mathbf{B}_\phi^\mathrm{T} \mathbf{B}_\phi \right) \mathrm{d}V, \tag{19}$$

$$\mathbf{K}_{u\phi} = \int_V \mathbf{N}_\phi^T \left(\frac{dg}{\mathrm{d}\phi} \frac{\mathrm{d}\psi_0^+}{\mathrm{d}\boldsymbol{\varepsilon}} \right) \mathbf{B}_u \mathrm{d}V \tag{20}$$

$$\mathbf{K}_{\phi u} = \int_V \mathbf{B}_u^\mathrm{T} \left(\frac{dg}{\mathrm{d}\phi} \frac{\mathrm{d}\psi_0^+}{\mathrm{d}\boldsymbol{\varepsilon}} \right) \mathbf{N}_\phi \mathrm{d}V \quad \text{for } \psi_0^+ > H_{i-1}, \tag{21}$$

$$\mathbf{K}_{u\phi} = 0 \quad \text{for } \psi_0^+ < H_{i-1}. \tag{22}$$

2.6 Solving Procedure

The staggered solution scheme is used in this work, similarly as in [24]. Firstly, a displacement iteration is performed to obtain a new displacement field in *the* k-th iteration of the ith increment:

$$\mathbf{v}_i^k = \mathbf{v}_i^{k-1} + \mathbf{K}_{uu}^{-1}\left(\mathbf{v}_i^{k-1}, \boldsymbol{\phi}_i^{k-1}\right) \mathbf{R}_u\left(\mathbf{v}_i^{k-1}, \boldsymbol{\phi}_i^{k-1}\right) \tag{23}$$

Using the newly obtained displacement field, a new phase-field is calculated as:

$$\boldsymbol{\phi}_i^k = \boldsymbol{\phi}_i^{k-1} + \mathrm{K}_{\phi\phi}^{-1}\left(\mathbf{v}_i^k, \boldsymbol{\phi}_i^{k-1}\right) \cdot \mathrm{R}_\phi\left(\mathbf{v}_i^k, \boldsymbol{\phi}_i^{k-1}\right) \tag{24}$$

After the phase-field iteration for $k > 1$ (in the first set of iterations the convergence check is skipped), convergence criteria are checked. The convergence criteria are defined as follows:

$$\left\| \boldsymbol{\phi}_i^k - \boldsymbol{\phi}_i^{k-1} \right\|_\infty < f_1, \tag{25}$$

$$\frac{\Delta \Psi_{e,i}^k}{\Delta \Psi_{e,i}^1} < f_2, \tag{26}$$

where $\Delta\Psi_{e,i}^{k}$ is change of deformational energy during displacement calculation in k-th iteration of i-increment. In this work $f_1 = f_2 = 0.001$ is used.

3 Remeshing Procedure

In this section, the proposed remeshing procedure is described. Remeshing is done after each phase-field iteration. In the first step of the remeshing, the algorithm loops between active finite elements and checks if a refinement criterion is satisfied. If the criterion is satisfied, the longest edge of the element is marked for splitting. Also, in a recursive manner, it is checked if that edge is the neighboring element's longest edge. If it is not, the longest edge of neighboring element is marked for splitting too (see Fig. 1). Following this rule will lead to a fine mesh without distortion. After edge marking, element splitting is performed by looping across all elements and splitting the elements into two new elements if any of the edges is marked for splitting. During element splitting, new nodes are defined on both the split edge and the newly defined edge. To prevent the duplication of nodes, the nodes are related to the edge and it is unique to the edge. An element is firstly split by splitting the longest edge

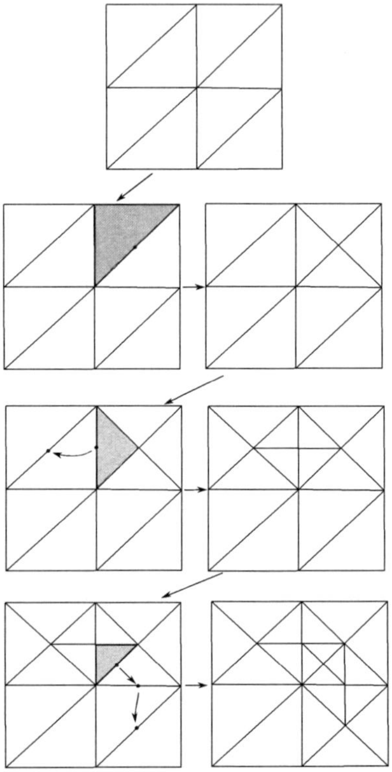

Fig. 1 Illustration of mesh refinement. Grey colored elements are the ones that satisfy refinement conditions

first. New/children elements are related to old/parent elements and those relations are memorized as it enables mesh coarsening. The children elements of one parent element are referred to each other as twin elements. During splitting, nodal variables of new nodes are interpolated using existing nodal variables and shape functions of the split element. Integration point data such as the history field are mapped by approximating data from the split finite element with a quadratic polynomial. After splitting, the parent element is deactivated, and it does not participate in the analysis.

Mesh coarsening is done similarly to refinement. Mesh hierarchy is utilized, and coarsening is performed by merging two twin elements. Hanging nodes are avoided by splitting edges with potential hanging nodes, while following the rule of splitting the longest edge first.

As one remeshing cycle can lead to only one level of refinement or coarsening, remeshing cycles are repeated until sufficient remeshing is performed or until the maximum number of remeshing cycles allowed in one iteration are done. In this work, up to 3 cycles are done.

Following refinement criteria are used, i.e., remeshing is triggered if any of these criteria is satisfied:

- phase-field criterion: $h > \frac{l}{k_1}\frac{1}{\phi^2+k_2} \& \phi_{\max} > \phi_0$,
- deformation energy density criterion: $h > \frac{l}{k_1} \quad \& \quad \psi_{max} > \frac{k_3 G_c}{2l}$,
- phase-field gradient criterion: $\phi_{max} - \phi_{min} > k_4$.

In the above expressions, h is the length of the longest edge of finite element, k_1 is the factor which defines how many elements per length-scale are necessary, k_2 is a small user-defined numerical parameter, k_3 is the factor which determines the deformational energy density which will trigger remeshing, and k_4 is the maximal allowed change in the phase-field in an element. Note that ψ_{max} is the maximum degraded positive part of deformation energy of all integration points in finite element. Also,ϕ_{max} and ϕ_{min} are maximum and minimum nodal phase-field values of the element. We argue that the deformation energy criterion would ensure accurate crack growth, as it will ensure sufficient mesh density around the crack tip and in the initiation zone prior to crack initiation. The phase-field and phase-field gradient criteria would ensure sufficient mesh density on crack parts distant to the crack tip. Also, an effective stress error criterion is used to ensure the accuracy of stress and displacement fields. I In this work $k_1 = 2$, $k_2 = 0.0001$, $k_3 = 0.15$ and $k_4 = 0.4$ are used. Coarsening criteria are similar to refinement criteria, with the only difference being stricter. Also, to trigger coarsening, all coarsening criteria have to be satisfied.

4 Results

In this section some relevant results are presented. The single-edge notched specimen example has been studied. The example is described by Fig. 2. The material

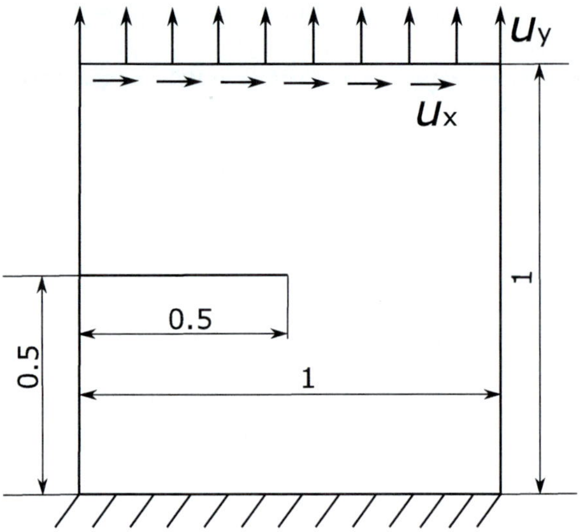

Fig. 2 Description of studied example

parameters are: the Young's modulus $E = 210000$, Poisson's rate $\upsilon = 0.3$, critical energy release rate $G_c = 2.7$ and length-scale $l = 0.0075$. Multiple cases have been studied including two types of boundary conditions: the tensile case ($u_x = 0, u_y = 0.0075$), and the pure shear case $u_x = 0.015, u_y = 0$. Loading is done in 50 uniform increments for the tensile case, and in 30 uniform increments for the shear case.

Multiple different meshes have been studied. The considered starting meshes are shown in Fig. 3. The first mesh is a structured mesh consisting of small numbers of the equally sized triangle elements (*mesh 1*). To test the robustness of the remeshing algorithm one unstructured mesh is used (*mesh 2*). Two dense meshes made with previous knowledge of crack paths are used without adaptive refinement and are used to obtain referent solutions for assessing the accuracy and computational efficiency of adaptive algorithm. The dense meshes consist of 22,357 elements for the tensile case and 38,015 elements for the shear case.

In contrast to a large number of elements which is necessary for accurate results in the case without adaptivity, with the proposed adaptive remeshing algorithm the final number of elements is kept small. For reference, the final mesh constructed by the remeshing algorithm in case of mesh 1 for the tensile case consists of 5140 active elements and 5658 active elements for the shear case. The final meshes constructed by the remeshing algorithm are shown in Figs. 4 and 5.

The resulting crack paths shown in Figs. 6 and 7. It is well known that phase-field results can be influenced by the topology of mesh, as the effective fracture toughness is influenced by the mesh size. We observe very small difference in crack path between both starting meshes with adaptivity (mesh 1 and 2) and dense mesh without adaptivity (mesh 3). Such result is a proof of algorithm's robustness.

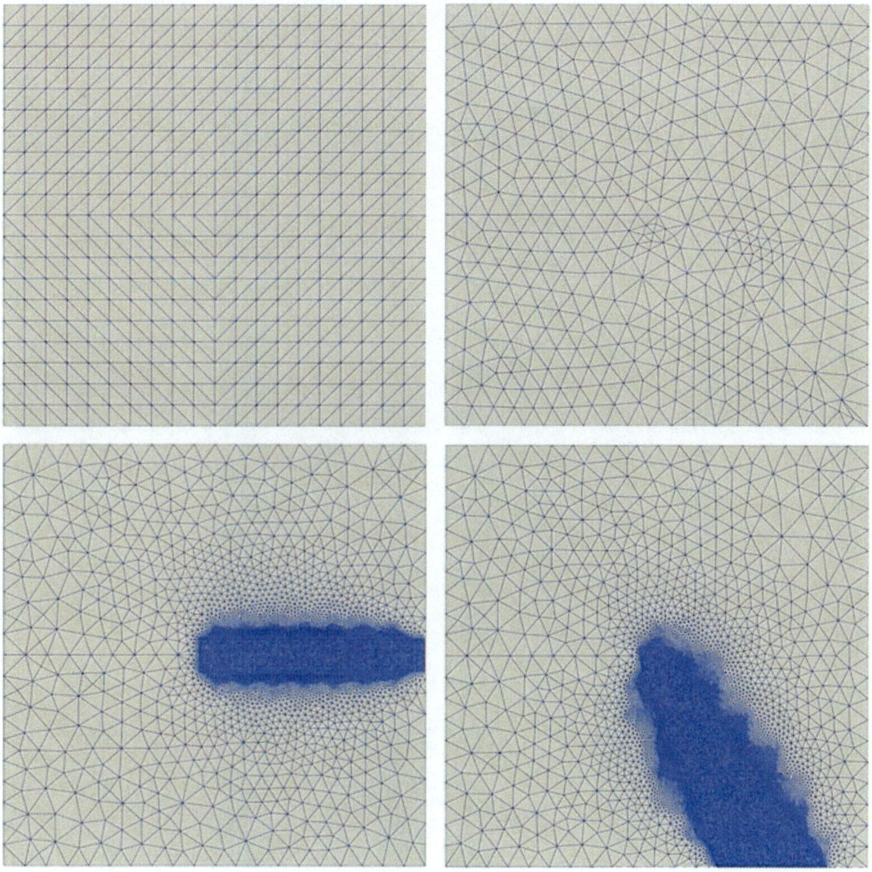

Fig. 3 Starting meshes: *mesh 1* - structured (top-left), *mesh 2* - unstructured (top-right), *mesh 3* - a dense mesh for the tensile case without adaptivity (bottom-left), *mesh 4* - a dense mesh for shear case without adaptive remeshing (bottom-right)

A further proof of remeshing algorithms robustness is visible from the force–displacement diagrams shown in Figs. 8 and 9. A very small error is visible after the crack initiation phase. Most importantly, crack initiation is described well, and there is no delay in crack growth in comparison to dense mesh without adaptivity. Proper remeshing prior to crack initiation is mostly influenced by the deformation energy density criterion and the stress error criterion.

Computational times are shown in Table 1. It is shown that significant computational time savings can be obtained in comparison to a dense mesh without adaptivity. It demonstrates the computational efficiency of the remeshing algorithm. Also, it suggests that the remeshing procedure is computationally undemanding in comparison to the time required for the numerical integration of the stiffness matrix and residual vectors and solving the system of equations. Note that computational time

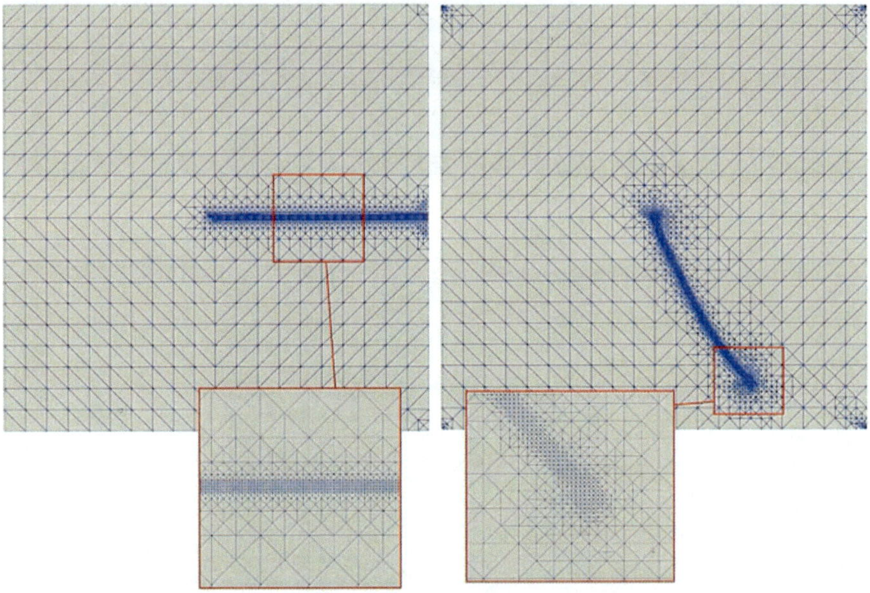

Fig. 4 Final meshes constructed by the remeshing algorithm from mesh 1

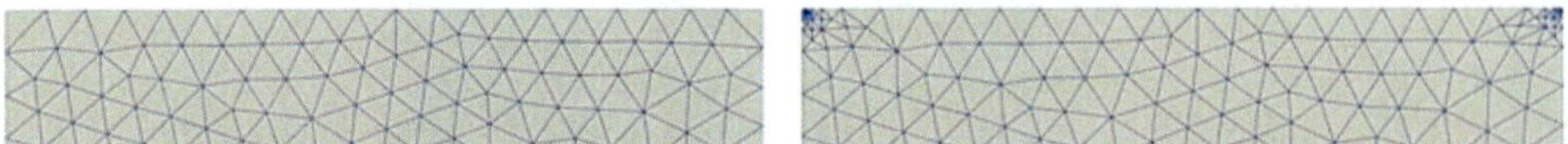

Fig. 5 Final meshes constructed by remeshing algorithm from mesh 2

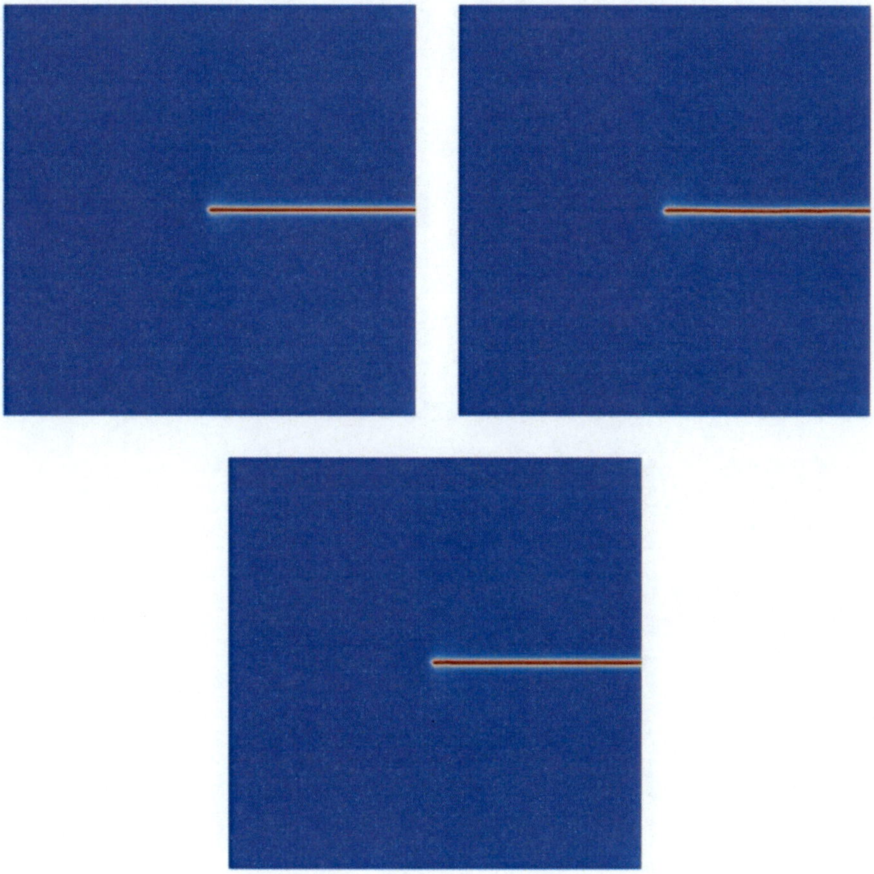

Fig. 6 Results obtained for tensile case for mesh 1 (top-left), mesh 2 (top-right) and dense mesh (bottom)

saving is an unobjective value, since it depends on a starting mesh. Usually, for simple cases crack paths can be assumed, and appropriate coarser meshes can be constructed prior the simulation. For complex geometries with unpredictable crack paths, a proper mesh would unavoidably be extremely large, and hence the computation time savings due to adaptive remeshing would be potentially large. Finally, it should be noted that the number of iterations is similar for all considered meshes and its impact on the total computational time is insignificant.

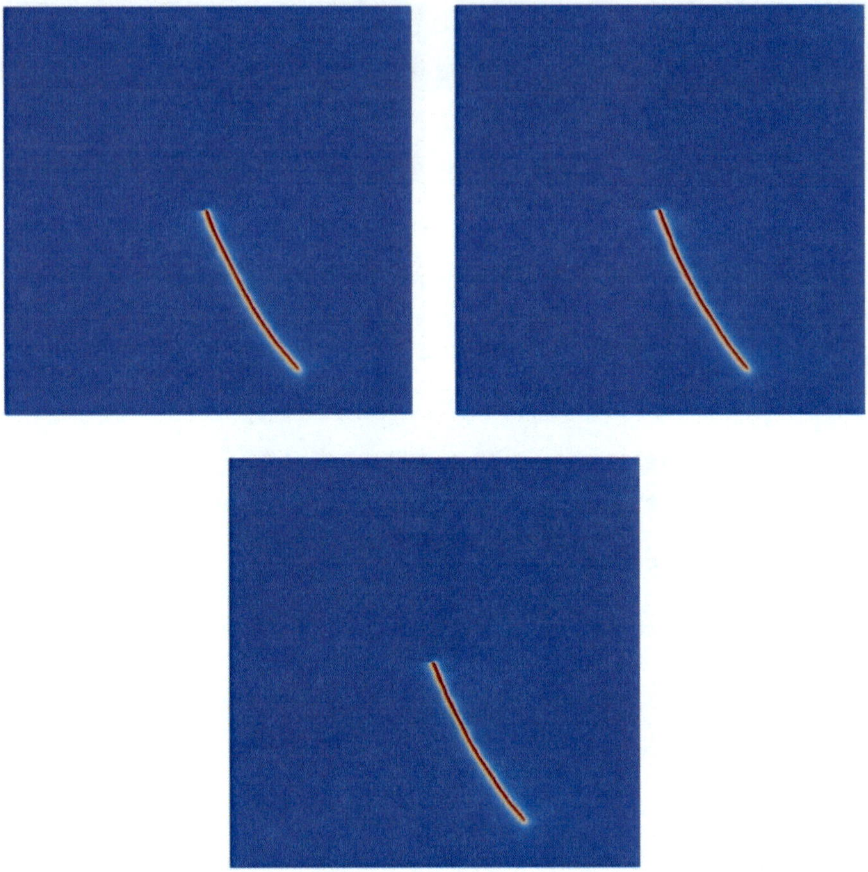

Fig. 7 Results obtained for the shear case: for mesh 1 (top-left), mesh 2 (top-right) and dense mesh (bottom)

5 Conclusion

In this paper, a simple adaptive remeshing procedure is presented. Remeshing is based on physical criteria and procedure is based on triangle splitting. The algorithm shows robustness and efficiency and the remeshing procedure has shown to be computational undemanding. Significant time savings have been obtained in comparison to the referent dense meshes without adaptivity, but it is important to note that such conclusion cannot be completely objective since computational savings depend on a starting mesh. Still, it is obvious from the numerical study that time savings can be obtained by using the proposed adaptive algorithm even for simple problems. Therefore, the proposed adaptive strategy demonstrates its potential for efficiently solving the problems with complex geometries, as it is capable of refining mesh prior to crack initiation.

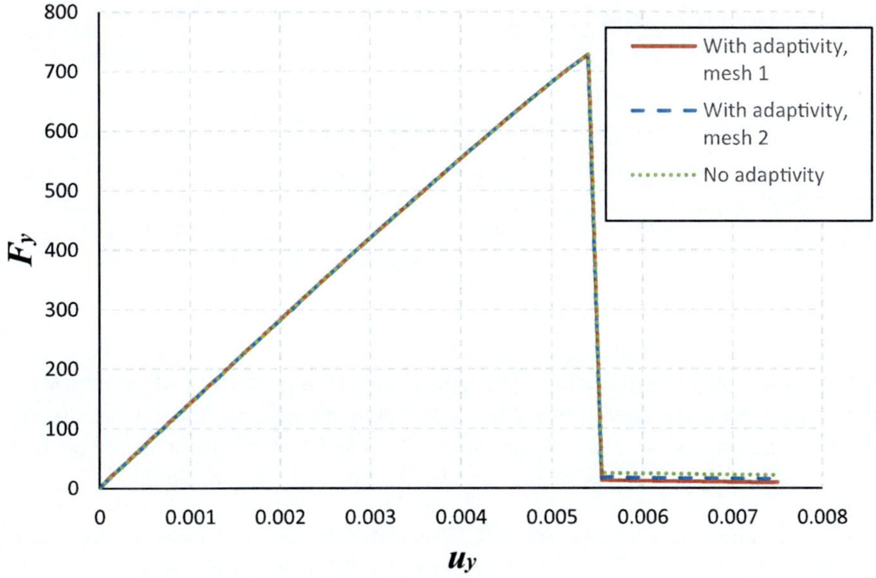

Fig. 8 Force displacement diagram for tensile case

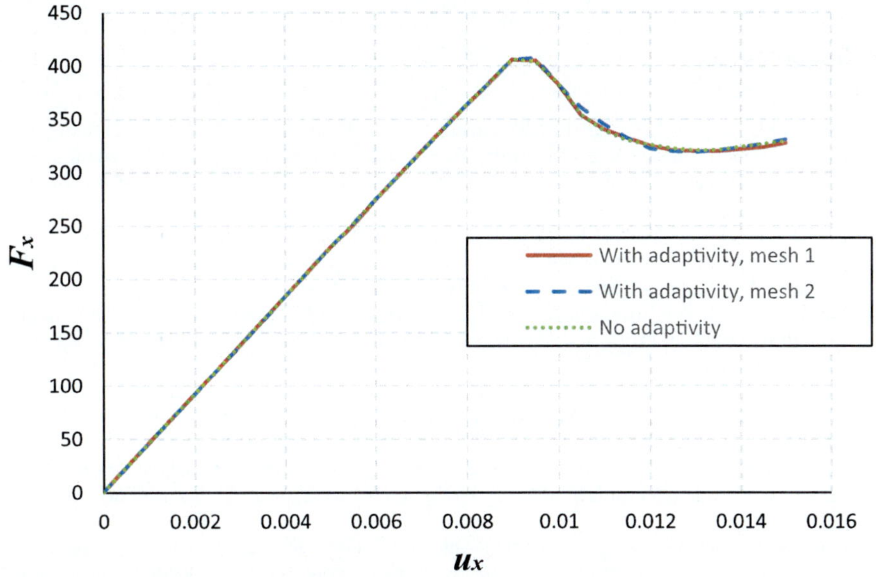

Fig. 9 Force displacement diagram for shear case

Table 1 Comparison of computational times

	Total time (s)				Computational time savings
	Adaptive remeshing			Dense mesh, no adaptivity	
	Computation time	Remeshing time	Sum		
Tensile, mesh 1	718.77	245.57	964.34	5625.41	82.85%
Tensile, mesh 2	974.93	349.94	1324.87		76.44%
Shear, mesh 1	5066.59	980.50	6047.09	59,329.95	88.39%
Shear, mesh 2	5690.50	1196.63	6887.13		89.80%

The future work will include the upgrading of the existing algorithm with anisotropic remeshing capabilities. Also, a further investigation of the influence of the algorithm parameters will be investigated. Additionally, an extension of the present work to different phenomena such as ductile fracture will be considered.

Acknowledgements This work has been fully supported by the Croatian *Ministry of Science and Education* under the project "Protection of Structural Integrity in Energy and Transport" (ZaCjel, KK.01.1.1.04.0056).

References

1. Borden MJ, Hughes TJR, Landis CM, Anvari A, Lee IJ (2017) A phase-field formulation for fracture in ductile materials: Finite deformation balance law derivation, plastic degradation, and stress triaxiality effects (vol 312, p 130, 2016). Comput Methods Appl Mech Eng 324:712–713. https://doi.org/10.1016/j.cma.2017.06.023
2. Bleyer J, Roux-Langlois C, Molinari J-F (2017) Dynamic crack propagation with a variational phase-field model: limiting speed, crack branching and velocity-toughening mechanisms. Int J Fract 204:1–22. https://doi.org/10.1007/s10704-016-0163-1
3. Seleš K, Aldakheel F, Tonkovic Z, Sorić J, and Wriggers P (2021) A general phase-field model for fatigue failure in brittle and ductile solids. Comput Mech 67:1431–1452. https://doi.org/10.1007/s00466-021-01996-5
4. Burke S, Ortner C, Suli E (2010) An adaptive finite element approximation of a variational model of brittle fracture. SIAM J Numer Anal 48(3):980–1012. https://doi.org/10.1137/080741033
5. Artina M, Fornasier M, Micheletti S, Perotto S (2015) Anisotropic mesh adaptation for crack detection in brittle materials. SIAM J Sci Comput 37(4):B633–B659. https://doi.org/10.1137/140970495
6. Ferro N, Micheletti S, Perotto S (2018) Anisotropic mesh adaptation for crack propagation induced by a thermal shock in 2D. Comput Methods Appl Mech Eng 331:138–158. https://doi.org/10.1016/j.cma.2017.11.024
7. Micheletti S, Perotto S, Signorini M (2018) Anisotropic mesh adaptation for the generalized Ambrosio-Tortorelli functional with application to brittle fracture. Comput Math Appl 75(6):2134–2152. https://doi.org/10.1016/j.camwa.2017.08.009
8. Heister T, Wheeler MF, Wick T (2015) A primal-dual active set method and predictor-corrector mesh adaptivity for computing fracture propagation using a phase-field approach. Comput Methods Appl Mech Eng 290:466–495. https://doi.org/10.1016/j.cma.2015.03.009

9. Mang K, Walloth M, Wick T, Wollner W (2019) Mesh adaptivity for quasi-static phase-field fractures based on a residual-type a posteriori error estimator. GAMM-Mitteilungen 43. https://doi.org/10.1002/gamm.202000003.
10. Klinsmann M, Rosato D, Kamlah M, McMeeking RM (2015) An assessment of the phase field formulation for crack growth. Comput Methods Appl Mech Eng 294:313–330. https://doi.org/10.1016/j.cma.2015.06.009
11. Hirshikesh, Jansari C, Kannan K, Annabattula RK, Natarajan S (2019) Adaptive phase field method for quasi-static brittle fracture using a recovery based error indicator and quadtree decomposition. Eng Fract Mech 220:19. https://doi.org/10.1016/j.engfracmech.2019.106599
12. Eldahshan H, Bouchard P-O, Alves J, Perchat E, Munoz DP (2021) Phase field modeling of ductile fracture at large plastic strains using adaptive isotropic remeshing. Comput Mech. https://doi.org/10.1007/s00466-020-01962-7
13. Badnava H, Msekh MA, Etemadi E, Rabczuk T (2018) An h-adaptive thermo-mechanical phase field model for fracture. Finite Elem Anal Des 138:31–47. https://doi.org/10.1016/j.finel.2017.09.003
14. Tian FC, Tang XL, Xu TY, Yang JS, Li LB (2019) A hybrid adaptive finite element phase-field method for quasi-static and dynamic brittle fracture. Int J Numer Meth Eng 120(9):1108–1125. https://doi.org/10.1002/nme.6172
15. Goswami S, Anitescu C, Rabczuk T (2019) Adaptive phase field analysis with dual hierarchical meshes for brittle fracture. Eng Fract Mech 2:218. https://doi.org/10.1016/j.engfracmech.2019.106608
16. Hussein A, Hudobivnik B, Wriggers P (2020) A combined adaptive phase field and discrete cutting method for the prediction of crack paths. Comput Methods Appl Mech Eng 372. https://doi.org/10.1016/j.cma.2020.113329
17. Francfort GA, Marigo JJ (1998) Revisiting brittle fracture as an energy minimization problem. J Mech Phys Solids 46(8):1319–1342. https://doi.org/10.1016/s0022-5096(98)00034-9
18. Griffith AA, Taylor GIVI (1921) The phenomena of rupture and flow in solids. Philos Trans R Soc Lond Ser A, Contain Pap Math Phys Character 221(582–593):163–198. https://doi.org/10.1098/rsta.1921.0006
19. Bourdin B, Francfort GA, Marigo JJ (2000) Numerical experiments in revisited brittle fracture. J Mech Phys Solids 48(4):797–826. https://doi.org/10.1016/s0022-5096(99)00028-9
20. Wu J-Y, Nguyen VP, Nguyen CT, Sutula D, Sinaie S, Bordas SPA (2020) Phase-field modeling of fracture. In: Bordas SPA, Balint DS (eds) Advances in applied mechanics. Elsevier, pp 1–183
21. Miehe C, Welschinger F, Hofacker M (2010) Thermodynamically consistent phase-field models of fracture: variational principles and multi-field FE implementations. Int J Numer Meth Eng 83(10):1273–1311. https://doi.org/10.1002/nme.2861
22. Miehe C, Hofacker M, Welschinger F (2010) A phase field model for rate-independent crack propagation: robust algorithmic implementation based on operator splits. Comput Methods Appl Mech Eng 199(45–48):2765–2778. https://doi.org/10.1016/j.cma.2010.04.011
23. De Lorenzis L, Gerasimov T (2020) Numerical implementation of phase-field models of brittle fracture. In: De Lorenzis L, Düster A (eds) Modeling in engineering using innovative numerical methods for solids and fluids. Springer International Publishing, Cham, pp 75–101
24. Seles K, Lesicar T, Tonkovic Z, Soric J (2019) A residual control staggered solution scheme for the phase-field modeling of brittle fracture. Eng Fract Mech 205:370–386. https://doi.org/10.1016/j.engfracmech.2018.09.027

Application of Multilayer Perceptron Neural Network for Damage Detection in Rectangular Laminated Composite Plates Based on Vibrational Analysis

Morteza Saadatmorad, Ramazan-Ali Jafari-Talookolaei, Mohammad-Hadi Pashaei, Samir Khatir, and Magd Abdel Wahab

Abstract In this paper, a new model is proposed to detect damages severities in rectangular laminated composite plates using a type of neural network called multilayer perceptron (MLP) and the finite element method (FEM). First, experimental data are produced by using FEM, and then the MLP network is applied to identify damages' severities in single damage scenarios. The damages' severities and the first five natural frequencies of rectangular laminated composite plates are assumed as outputs and inputs of the network, respectively. According to these data, the MLP neural network uses a hyperbolic tangent sigmoid activation function to be trained and predict the damages' severities in rectangular laminated composite plates. Our findings demonstrate that the MLP neural network model can identify the severity of every single damage scenario in rectangular laminated composite plates with a high precision ($R > 0.9$) by having experimental natural frequencies.

Keywords Damaged laminated composites · Multilayer perceptron · Damaged structures · Finite element analysis

1 Introduction

In recent years, composite materials have been widely used to construct various structures due to their high strength and stiffness, lightness, etc. [1]. Composite materials are also used in many high-speed applications such as automotive engineering, aerospace engineering, etc. [2]. Thus, they are subjected to serve impacts and damages that can significantly change their structural behavior [3]. Therefore, damage detection has attracted many researchers' attention to develop novel and useful methods for monitoring structures' health.

M. Saadatmorad · R.-A. Jafari-Talookolaei (✉) · M.-H. Pashaei
Babol Noshirvani University of Technology, Mazandaran province, Babol, Iran
e-mail: ra.jafari@nit.ac.ir

S. Khatir · M. Abdel Wahab
Soete Laboratory, Faculty of Engineering and Architecture, Ghent University, Technologiepark Zwijnaarde 903, B-9052 Zwijnaarde, Belgium

M. Abdel Wahab (ed.), *Proceedings of the 2nd International Conference on Structural Damage Modelling and Assessment*, Lecture Notes in Civil Engineering 204,
https://doi.org/10.1007/978-981-16-7216-3_13

Generally, analysis of damaged structures is divided into two significant types: forward analysis and inverse analysis. The forward analysis's objective is to obtain the natural frequencies or dynamic characteristics of a vibrational structure. Therefore, the severity and location of damages are specified in the forward analysis [4]. In contrast, the inverse analysis aims to find the severity and location of damages by having a vibrational structure's natural frequencies or dynamic characteristics [5]. In recent years, many methodologies have been proposed for the inverse analysis of vibrational structures. The artificial intelligence-based inverse analysis is one of the most recent methodologies. These methodologies use soft computing tools such as evolutional algorithms and machine learning.

Along with the artificial intelligence-based inverse analysis, Dinh-Cong et al. applied a Jaya algorithm to detect plate structures' damage [6]. Gomes et al. used the sunflower optimization technique to detect damage in composite plates [7]. Khatir et al. [1] studied damage identification in rectangular laminated composite plates. A damage index called local frequency change ratio was introduced to assess damages in a three-layer laminated composite. Findings showed that this index could detect damage location with high accuracy. Many other studies of optimization-based damage detection have been presented in recent years. For example, Barman et al. [8] compared particle swarm optimization technique with ant colony optimization to identify damages in space truss and plane structures. Gerist et al. [9] applied the imperialist competitive algorithm to identify damage via a novel mode shapes-based objective function.

In addition to the optimization-based approach in inverse analysis, the artificial neural network-based approach is also most commonly used in literature. Artificial neural networks (ANNs) are beneficial for predicting damage in structures [10]. Liu et al. [11] applied normalized flexibility vectors as the input of an ANN to identify the damages' severity and location. Nasiri et al. used ANN and modal analysis to identify the severity of a common type of inter-laminar damage called delamination [12]. Yam et al. proposed the use of neural networks and wavelet transform for damage identification of composite structures. Khatir et al. [13] propose a powerful yet simple, novel method based on ANNs combined with an optimization algorithm called Particle Swarm Optimization for quantifying damage in composite plates via an index called Cornwell indicator. Yu et al. [14] presented a novel online approach based on ANNs to identify laminated composite shells' damages partially filled with fluid.

According to our literature review, ANNs have been used in various strategies to detect damages in structures. However, this paper presents an efficient multi-layer perceptron neural network for the first time to predict the damages' severity in rectangular laminated composite plates by updating the finite element method.

2 Methodology

2.1 Mathematical Formulation

Consider the plates shown in Fig. 1. The figure shows two plates one plate is intact (a) and the other is a damaged (b). Mathematical expression for the dynamic equation of a rectangular laminated composite plate is obtained using the FEM. For this, a rectangular laminated composite plate as a continuous system is discretized as finite elements. Local stiffness and mass matrices for each element are derived based on the first shear deformation theory (FSDT). Finally, global stiffness and mass matrices are obtained to express the equation of motion.

Constitutive equations

According to the FSDT, displacement fields for each layer of the rectangular laminated composite plate are presented as [12]:

$$\begin{aligned} u(x, y, z) &= u_0(x, y) + z\psi_x(x, y) \\ v(x, y, z) &= v_0(x, y) + z\psi_y(x, y) \\ w(x, y, z) &= w_0(x, y) \end{aligned} \tag{1}$$

where, u_i and $v_i (i = 0, 1, 2, 3)$ are the in-plane displacements of each layer of the rectangular laminated composite plate, as well as $w_j (j = 0, 1, 2)$ represent vertical displacements of each layer of the plate. ψ_x and ψ_y are rotations about x and y, respectively.

By having these displacements, generalized strains are obtained to express the constitutive equations as follows:

Fig. 1 Geometrical model of two rectangular laminated composite plates: **a** intact plate, **b** damaged plate

$$\begin{Bmatrix} N_x \\ N_y \\ N_{xy} \\ M_x \\ M_y \\ M_{xy} \end{Bmatrix} = \begin{bmatrix} A_{11} & A_{12} & A_{16} & B_{11} & B_{12} & B_{16} \\ A_{12} & A_{22} & A_{26} & B_{12} & B_{22} & B_{26} \\ A_{16} & A_{26} & A_{66} & B_{16} & B_{26} & B_{66} \\ B_{11} & B_{12} & B_{16} & D_{11} & D_{12} & D_{16} \\ B_{12} & B_{22} & B_{26} & D_{12} & D_{22} & D_{26} \\ B_{16} & B_{26} & B_{66} & D_{16} & D_{26} & D_{66} \end{bmatrix} \begin{Bmatrix} \varepsilon_x^0 \\ \varepsilon_y^0 \\ \gamma_{xy}^0 \\ \kappa_x \\ \kappa_y \\ \kappa_{xy} \end{Bmatrix} \tag{2}$$

$$\begin{Bmatrix} Q_{yz} \\ Q_{xz} \end{Bmatrix} = \begin{bmatrix} A_{44} & A_{45} \\ A_{45} & A_{55} \end{bmatrix} \begin{Bmatrix} \gamma_{yz} \\ \gamma_{xz} \end{Bmatrix} \tag{3}$$

where, A_{ij},B_{ij}, and $D_{ij}(i, j = 1, 2, \ldots, 6)$ are the extension, bending–extension coupling, and bending stiffness's, respectively. $A_{ij}(i, j = 4, 5)$ is the transverse shear stiffness's. A_{ij}, B_{ij}, and D_{ij} can be given as [12]:

$$A_{ij} = \sum_{k=1}^{N} \left(\overline{Q}_{ij}\right)_k (z_k - z_{k-1}) \quad \text{i, j} = 1, 2, 6 \tag{4}$$

$$B_{ij} = \frac{1}{2} \sum_{k=1}^{N} \left(\overline{Q}_{ij}\right)_k \left(z_k^2 - z_{k-1}^2\right) \quad \text{i, j} = 1, 2, 6 \tag{5}$$

$$D_{ij} = \frac{1}{3} \sum_{k=1}^{N} \left(\overline{Q}_{ij}\right)_k \left(z_k^3 - z_{k-1}^3\right) \quad \text{i, j} = 1, 2, 6 \tag{6}$$

$$A_{ij} = k_{ij} \sum_{k=1}^{N} \left(\overline{Q}_{ij}\right)_k (z_k - z_{k-1}) \quad \text{i, j} = 4, 5 \tag{7}$$

where, $\overline{Q}_{ij}$ is the transformed stiffness coefficients of the laminated composite layers, N is the total number of layers in the laminated composite, z_k indicates the distance of the surface of the kth layer to the origin. In (7), k_{ij} is the shear correction coefficient $(k_{ij} = 0.833)$.

By the above assumptions, the considered rectangular laminated composite plate is converted into one lamina that the Eqs. (2) and (3) are governed on it. As a result, the Eqs. (4)-(7) present the equivalent stiffness's.

Development of finite element equations.

In this paper, a nine-node higher-order rectangular element is considered, as shown in Fig. 2.

Each node has five degrees of freedom (u_0, v_0,w_0,ψ_x, and ψ_y). Where the displacement field functions u_0, v_0,w_0,ψ_x, and ψ_y are written in terms of nodal displacement vector $\{d\}_{45\times1}$ and the shape functions N_{ψ_y}, N_{ψ_x}, N_w,N_v, and N_u. $\{d\}_{45\times1}$ is presented as: $\{d\}_{45\times1} = \left\{u_1, v_1, w_1, \psi_{x1}, \psi_{y1}, \ldots, u_9, v_9, w_9, \psi_{x9}, \psi_{y9}\right\}^T$.

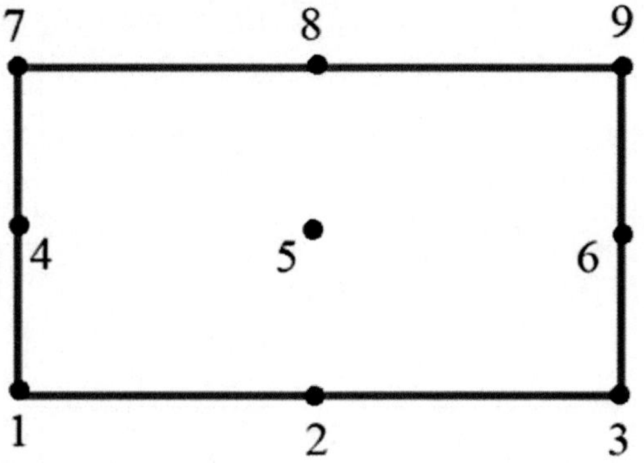

Fig. 2 The considered higher-order rectangular element with nine nodes

The Lagrange interpolation is used to find the value of each node's degree of freedom. To obtain local stiffness and mass matrices, first, the potential and kinetic energies of the considered rectangular laminated composite plate are obtained based on FSDT theory, then; rewrite them in terms of nodal displacements.

$$U_P = 0.5\int_0^a\int_0^b \big[A_{11}u_{0,x}^2 + A_{22}\upsilon_{0,y}^2 + A_{66}\big(u_{0,y}^2 + \upsilon_{0,x}^2 + 2u_{0,y}\upsilon_{0,x}\big) + D_{11}\psi_{x,x}^2 + D_{22}\psi_{y,y}^2 + D_{66}\big(\psi_{x,y}^2 + \psi_{y,x}^2 + 2\psi_{x,y}\psi_{y,x}\big) + A_{44}\big(\psi_y^2 + w_{0,y}^2 + 2\psi_y w_{0,y}\big) + A_{55}\big(\psi_x^2 + w_{0,x}^2 + 2\psi_x w_{0,x}\big) + 2A_{12}u_{0,x}\upsilon_{0,y} + 2A_{16}\big(u_{0,y} + \upsilon_{0,x}\big)u_{0,x} + 2B_{11}\psi_{x,x}u_{0,x} + 2B_{12}\psi_{y,y}u_{0,x} + 2B_{16}\big(\psi_{x,y} + \psi_{y,x}\big)u_{0,x} + 2A_{26}\big(u_{0,y} + \upsilon_{0,x}\big)\upsilon_{0,y} + 2B_{12}\psi_{x,x}\upsilon_{0,y} + 2B_{22}\psi_{y,y}\upsilon_{0,y} + 2B_{26}\big(\psi_{x,y} + \psi_{y,x}\big)\upsilon_{0,y} + 2B_{16}\psi_{x,x}\big(u_{0,y} + \upsilon_{0,x}\big) + 2B_{26}\psi_{y,y}\big(u_{0,y} + \upsilon_{0,x}\big) + 2B_{66}\big(\psi_{x,y} + \psi_{y,x}\big)\big(u_{0,y} + \upsilon_{0,x}\big) + 2D_{12}\psi_{y,y}\psi_{x,x} + 2D_{16}\big(\psi_{x,y} + \psi_{y,x}\big)\emptyset_{,x} + 2D_{26}\big(\psi_{x,y} + \psi_{y,x}\big)\psi_{y,y} + 2A_{45}\big(\psi_x + w_{0,x}\big)\big(\psi_y + w_{0,y}\big)\big]\mathrm{dydx} \tag{8}$$

It can be expressed in terms of local stiffness matrix $[k_e]$ as:

$$U_P = \frac{1}{2}\{d\}_{45\times 1}^T \cdot [k_e]_{45\times 45}\{d\}_{45\times 1} \tag{9}$$

As a result, the local stiffness matrix $[k_e]$ can be written as:

$$
\begin{aligned}
[k_e] = \int_0^a \int_0^b [\ & A_{11}\lfloor N_{u_{0,x}}\rfloor^T.\lfloor N_{u_{0,x}}\rfloor + A_{22}\lfloor N_{v_{0,y}}\rfloor^T.\lfloor N_{v_{0,y}}\rfloor + A_{66}(\lfloor N_{u_{0,y}}\rfloor^T.\lfloor N_{u_{0,y}}\rfloor + \\
& \lfloor N_{v_{0,x}}\rfloor^T.\lfloor N_{v_{0,x}}\rfloor + \lfloor N_{u_{0,y}}\rfloor^T.\lfloor N_{v_{0,x}}\rfloor + \lfloor N_{v_{0,x}}\rfloor^T.\lfloor N_{u_{0,y}}\rfloor A_{66}(\lfloor N_{u_{0,y}}\rfloor^T.\lfloor N_{u_{0,y}}\rfloor + \\
& \lfloor N_{v_{0,x}}\rfloor^T.\lfloor N_{v_{0,x}}\rfloor + \lfloor N_{u_{0,y}}\rfloor^T.\lfloor N_{v_{0,x}}\rfloor + \lfloor N_{v_{0,x}}\rfloor^T.\lfloor N_{u_{0,y}}\rfloor) + D_{11}\lfloor N_{\psi_{x,x}}\rfloor^T.\lfloor N_{\psi_{x,x}}\rfloor + \\
& D_{22}\lfloor N_{\psi_{y,y}}\rfloor^T.\lfloor N_{\psi_{y,y}}\rfloor + D_{66}\left(\lfloor N_{\psi_{x,y}}\rfloor^T.\lfloor N_{\psi_{x,y}}\rfloor + \lfloor N_{\psi_{y,x}}\rfloor^T.\lfloor N_{\psi_{y,x}}\rfloor + \right. \\
& \left.\lfloor N_{\psi_{xy}}\rfloor^T.\lfloor N_{\psi_{y,x}}\rfloor + \lfloor N_{\psi_{y,x}}\rfloor^T.\lfloor N_{\psi_{x,y}}\rfloor\right) + A_{44}\left(\lfloor N_{\psi_y}\rfloor^T.\lfloor N_{\psi_y}\rfloor + \lfloor N_{w_{0,y}}\rfloor^T.\lfloor N_{w_{0,y}}\rfloor + \right. \\
& \left.\lfloor N_{\psi_y}\rfloor^T.\lfloor N_{w_{0,y}}\rfloor + \lfloor N_{w_{0,y}}\rfloor^T.\lfloor N_{\psi_y}\rfloor\right) A_{44}\left(\lfloor N_{\psi_y}\rfloor^T.\lfloor N_{\psi_y}\rfloor + \lfloor N_{w_{0,y}}\rfloor^T.\lfloor N_{w_{0,y}}\rfloor + \right. \\
& \left.\lfloor N_{\psi_y}\rfloor^T.\lfloor N_{w_{0,y}}\rfloor + \lfloor N_{w_{0,y}}\rfloor^T.\lfloor N_{\psi_y}\rfloor\right) A_{44}\left(\lfloor N_{\psi_y}\rfloor^T.\lfloor N_{\psi_y}\rfloor + \lfloor N_{w_{0,y}}\rfloor^T.\lfloor N_{w_{0,y}}\rfloor + \right. \\
& \left.\lfloor N_{\psi_y}\rfloor^T.\lfloor N_{w_{0,y}}\rfloor + \lfloor N_{w_{0,y}}\rfloor^T.\lfloor N_{\psi_y}\rfloor\right) + A_{55}\left(\lfloor N_{\psi_x}\rfloor^T.\lfloor N_{\psi_x}\rfloor + \lfloor N_{w_{0,x}}\rfloor^T.\lfloor N_{w_{0,x}}\rfloor + \right. \\
& \left.\lfloor N_{\psi_x}\rfloor^T.\lfloor N_{w_{0,x}}\rfloor + \lfloor N_{w_{0,x}}\rfloor^T.\lfloor N_{\psi_x}\rfloor\right) + A_{12}(\lfloor N_{u_{0,x}}\rfloor^T.\lfloor N_{v_{0,y}}\rfloor + \lfloor N_{v_{0,y}}\rfloor^T.\lfloor N_{u_{0,x}}\rfloor + \\
& A_{16}(\lfloor N_{u_{0,y}}\rfloor^T.\lfloor N_{u_{0,x}}\rfloor + \lfloor N_{u_{0,x}}\rfloor^T.\lfloor N_{u_{0,y}}\rfloor + \lfloor N_{v_{0,x}}\rfloor^T.\lfloor N_{u_{0,x}}\rfloor + \lfloor N_{u_{0,x}}\rfloor^T.\lfloor N_{v_{0,x}}\rfloor + \\
& B_{11}\left(\lfloor N_{\psi_{x,x}}\rfloor^T.\lfloor N_{u_{0,x}}\rfloor + \lfloor N_{u_{0,x}}\rfloor^T.\lfloor N_{\psi_{x,x}}\rfloor\right) + B_{12}\left(\lfloor N_{\psi_{y,y}}\rfloor^T.\lfloor N_{u_{0,x}}\rfloor + \right. \\
& \left.\lfloor N_{u_{0,x}}\rfloor^T.\lfloor N_{\psi_{y,y}}\rfloor\right) B_{12}\left(\lfloor N_{\psi_{y,y}}\rfloor^T.\lfloor N_{u_{0,x}}\rfloor + \lfloor N_{u_{0,x}}\rfloor^T.\lfloor N_{\psi_{y,y}}\rfloor\right) + \\
& B_{16}(\lfloor N_{\psi_{x,y}}\rfloor^T.\lfloor N_{u_{0,x}}\rfloor + \lfloor N_{u_{0,x}}\rfloor^T.\lfloor N_{\psi_{x,y}}\rfloor + \lfloor N_{\psi_{y,x}}\rfloor^T.\lfloor N_{u_{0,x}}\rfloor + \lfloor N_{u_{0,x}}\rfloor^T.\lfloor N_{\psi_{y,x}}\rfloor) + \\
& A_{26}(\lfloor N_{u_{0,y}}\rfloor^T.\lfloor N_{v_{0,y}}\rfloor + \lfloor N_{v_{0,y}}\rfloor^T.\lfloor N_{u_{0,y}}\rfloor + \lfloor N_{v_{0,x}}\rfloor^T.\lfloor N_{v_{0,y}}\rfloor + \\
& \lfloor N_{v_{0,y}}\rfloor^T.\lfloor N_{v_{0,x}}\rfloor) + B_{12}\left(\lfloor N_{\psi_{x,x}}\rfloor^T.\lfloor N_{v_{0,y}}\rfloor + \lfloor N_{v_{0,y}}\rfloor^T.\lfloor N_{\psi_{x,x}}\rfloor\right) + \\
& B_{26}\left(\lfloor N_{\psi_{x,y}}\rfloor^T.\lfloor N_{v_{0,y}}\rfloor + \lfloor N_{v_{0,y}}\rfloor^T.\lfloor N_{\psi_{x,y}}\rfloor + \lfloor N_{\psi_{y,x}}\rfloor^T.\lfloor N_{v_{0,y}}\rfloor + \lfloor N_{v_{0,y}}\rfloor^T.\lfloor N_{\psi_{y,x}}\rfloor\right) + \\
& B_{26}\left(\lfloor N_{\psi_{x,y}}\rfloor^T.\lfloor N_{v_{0,y}}\rfloor + \lfloor N_{v_{0,y}}\rfloor^T.\lfloor N_{\psi_{x,y}}\rfloor + \lfloor N_{\psi_{y,x}}\rfloor^T.\lfloor N_{v_{0,y}}\rfloor + \lfloor N_{v_{0,y}}\rfloor^T.\lfloor N_{\psi_{y,x}}\rfloor\right) + \\
& + B_{16}(\lfloor N_{\psi_{x,x}}\rfloor^T.\lfloor N_{u_{0,y}}\rfloor + \lfloor N_{u_{0,y}}\rfloor^T.\lfloor N_{\psi_{x,x}}\rfloor + \lfloor N_{\psi_{x,x}}\rfloor^T +.\lfloor N_{v_{0,x}}\rfloor + \\
& \lfloor N_{v_{0,y}}\rfloor^T.\lfloor N_{\psi_{x,y}}\rfloor) + \\
& B_{26}(\lfloor N_{\psi_{y,y}}\rfloor^T.\lfloor N_{u_{0,y}}\rfloor + \lfloor N_{u_{0,y}}\rfloor^T.\lfloor N_{\psi_{y,y}}\rfloor + \lfloor N_{\psi_{y,y}}\rfloor^T.\lfloor N_{v_{0,x}}\rfloor + \lfloor N_{v_{0,x}}\rfloor^T.\lfloor N_{\psi_{y,y}}\rfloor) + \\
& B_{66}(\lfloor N_{\psi_{x,y}}\rfloor^T.\lfloor N_{u_{0,y}}\rfloor + \lfloor N_{u_{0,y}}\rfloor^T.\lfloor N_{\psi_{x,y}}\rfloor + \lfloor N_{\psi_{x,y}}\rfloor^T.\lfloor N_{v_{0,x}}\rfloor + \\
& \lfloor N_{v_{0,x}}\rfloor^T.\lfloor N_{\psi_{x,y}}\rfloor + \lfloor N_{\psi_{y,x}}\rfloor^T.\lfloor N_{u_{0,y}}\rfloor + \lfloor N_{u_{0,y}}\rfloor^T.\lfloor N_{\psi_{y,x}}\rfloor + \lfloor N_{\psi_{y,x}}\rfloor^T.\lfloor N_{v_{0,x}}\rfloor + \\
& \lfloor N_{v_{0,x}}\rfloor^T.\lfloor N_{\psi_{y,x}}\rfloor) + D_{26}(\lfloor N_{\psi_{x,y}}\rfloor^T.\lfloor N_{\psi_{y,y}}\rfloor + \lfloor N_{\psi_{y,y}}\rfloor^T.\lfloor N_{\psi_{x,y}}\rfloor + \lfloor N_{\psi_{y,x}}\rfloor^T.\lfloor N_{\psi_{y,y}}\rfloor) + \\
& A_{45}(\lfloor N_{\psi_x}\rfloor^T.\lfloor N_{\psi_y}\rfloor + \lfloor N_{\psi_y}\rfloor^T.\lfloor N_{\psi_x}\rfloor + \lfloor N_{\psi_x}\rfloor^T.\lfloor N_{w_{0,y}}\rfloor + \lfloor N_{w_{0,y}}\rfloor^T.\lfloor N_{\psi_x}\rfloor + \lfloor N_{w_{0,x}}\rfloor^T.\lfloor N_{\psi_y}\rfloor \\
& + \lfloor N_{\psi_y}\rfloor^T.\lfloor N_{w_{0,x}}\rfloor + \lfloor N_{w_{0,x}}\rfloor^T.\lfloor N_{w_{0,y}}\rfloor + \lfloor N_{w_{0,y}}\rfloor^T.\lfloor N_{w_{0,x}}\rfloor)]\,\mathrm{dydx}
\end{aligned}
\tag{10}
$$

The kinetic energy of in one considered element of the rectangular laminated composite plate based on the FSDT is expressed as follows:

$$\mathrm{T} = \frac{1}{2}\int_0^a\int_0^b \left[I_0\left(u_{,t}^2 + v_t^2 + w_{,t}^2\right) + 2I_1\left(u_{,t}\psi_{x,t} + v_{,t}\psi_{y,t}\right) + I_2\left(\psi_{x,t}^2 + \psi_{y,t}^2\right)\right]dydx \tag{11}$$

where (I_0, I_1, I_2) are the mass moment of inertia:

$$(I_0, I_1, I_2) = \int_{-h/2}^{h/2} \rho(z)\left(1, z, z^2\right)dz \tag{12}$$

It can be expressed in terms of local mass matrix $[m_e]$ as:

$$\mathrm{T} = \frac{1}{2}\{d\}^T \cdot [m_e] \cdot \{d\} \tag{13}$$

where

$$\begin{aligned}[M_e] = \int_0^a\int_0^b &\Big[I_0\left(\lfloor N_u\rfloor^T \cdot \lfloor N_u\rfloor + \lfloor N_v\rfloor^T \cdot \lfloor N_v\rfloor + \lfloor N_w\rfloor^T \cdot \lfloor N_w\rfloor\right) \\ &+ I_1\left(\lfloor N_u\rfloor^T \cdot \lfloor N_{\psi_x}\rfloor + \lfloor N_{\psi_x}\rfloor^T \cdot \lfloor N_u\rfloor + \lfloor N_v\rfloor^T \cdot \lfloor N_{\psi_y}\rfloor + \lfloor N_{\psi_y}\rfloor^T \cdot \lfloor N_v\rfloor\right) \\ &+ I_2\left(\lfloor N_{\psi_x}\rfloor^T \cdot \lfloor N_{\psi_x}\rfloor + \lfloor N_{\psi_y}\rfloor^T \cdot \lfloor N_{\psi_y}\rfloor\right)\Big]dydx\end{aligned}$$

Governing equation of motion.

Global stiffness matrix [K] and global mass matrix $[M]$ are obtained by assembling local stiffness matrices $[k_e]$ of all elements and local mass matrices $[m_e]$ of all elements, respectively. Finally, the governing equation of motion based on FEM is expressed as follows:

$$[\mathrm{M}]\{\ddot{\Delta}\} + [\mathrm{K}]\{\Delta\} = 0 \tag{14}$$

where $\{\Delta\}$ includes the degrees of freedom for all nodes in system. By assuming $\{\Delta\} = (\Delta_0)e^{i\omega t}$ as well as $\lambda = \omega^2$, it can be written:

$$([\mathrm{K}] - \lambda[\mathrm{M}])\{\Delta_0\} = 0 \tag{15}$$

where ω denotes the natural frequency and $\{\Delta_0\}$ is the corresponding modes shapes to the natural frequencies. The non-zero solution of the system is calculated through the following equation using Matlab software.

$$\det([K] - \lambda[M]) = 0 \tag{16}$$

2.2 Multilayer Perceptron (MLP) Neural Network

In this study, a feed-forward artificial neural network called multilayer perceptron (MLP) is applied to identify the severity of damage in the single damaged rectangular laminated composite plates Fig. 3 indicates that the MLP consist of at least three layers of nodes: the hidden layer input layer and output layer nodes in output and hidden layers apply activation function [13].

$$input_k = \sum_{i,k=1}^{n,m} b_k + w_{ik} * x_i; \quad \mathrm{k} = (1 : \mathrm{m}); \mathrm{i} = (1 : \mathrm{n}) \tag{17}$$

where w_{ik} denotes weight for the ith neurons and the kth neurons, b_k indicates the bias ratio for hidden layers and inputs; x_i shows the output in i^{th} neurons of the input layer; m and n represent the number of the neurons in hidden and input layers, respectively. Also, $input_k$ indicates the input of kth neurons in the hidden layer.

Based on the following nonlinear transfer function, each neuron's $input_k$ is applied to compute the output of the jth neuron:

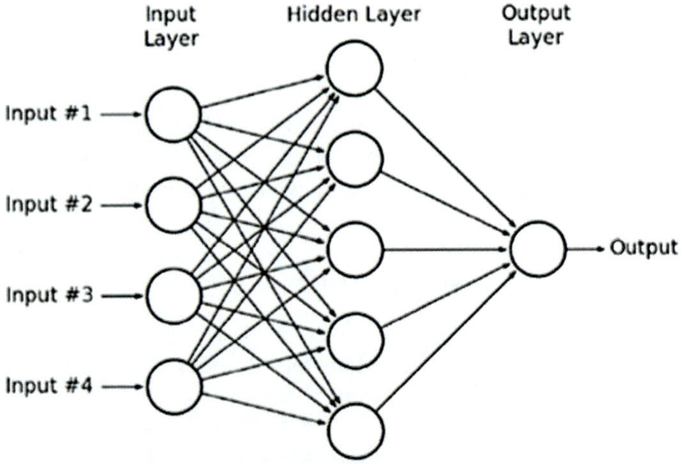

Fig. 3 The considered MLP [14]

$$ouput_j = \left(\frac{2}{1 + \frac{1}{e_q^{2input}}} \right) - 1 \tag{18}$$

This function is called the Hyperbolic Tangent Sigmoid activation function.

Also, in (18), $input_q$ denotes input of qth neuron in the output layer, and $ouput_j$ indicates j^{th} neuron of the output layer.

3 Proposed Model

This paper proposes a novel methodology to detect the severity of single-damage scenarios in rectangular laminated composite plates based on FEM and MLP-ANN. At first, a finite element model is developed for an intact rectangular laminated composite plate to obtain five first natural frequencies and their corresponding mode shapes.

Then, single damage is applied to the rectangular laminated composite plate to obtain five first natural frequencies and their corresponding mode shapes in a single-damage scenario and investigate the effect of single-damage in the rectangular laminated composite plate.

As seen in Fig. 4, a damage index called α_{rand} is introduced as the coefficient of the local stiffness matrix $[k_e]_{i,j}$ to represent the presence of a single-damage in the location of (i,j) in the rectangular laminated composite plate. The value of the damage index is between 0 to 1 (i.e., $0 < \alpha_{rand} < 1$) because if = 0, then there is no local element in the location(i,j), and if = 1, then there is no damage in the local element in the location of(i,j). 100 α_{rand} s are produces randomly.

Two hundred α_{rand} are produced randomly to be used as the coefficients of $[k_e]_{i,j}$; thus, two hundred corresponding natural frequencies ($\omega_{n/corr}$) are produced. Therefore, α_{rand} and $\omega_{n/corr}$ are considered as the inputs and outputs for our finite element model, respectively. Then, as shown in Fig. 4, α_{rand} and $\omega_{n/corr}$ are used as the outputs and inputs for our ANN-MPL network, respectively, to train and test its performance. When the regression factor(R) is more than 0.9, our ANN-MPL will be a high-performance network. Therefore, it can predict every single-damage severity by having experimental natural frequencies of the rectangular laminated composite plate.

3.1 Computational Examples

A two-layer square laminated composite plate $[0/\frac{\pi}{4}]$ with boundary conditions clamp-clamp-clamp-clamp (CCCC) is discretized into 121 finite elements to investigate the effects of single-damage on the rectangular laminated composite plate's modal characteristics. As mentioned, a nine-node higher-order rectangular element is used

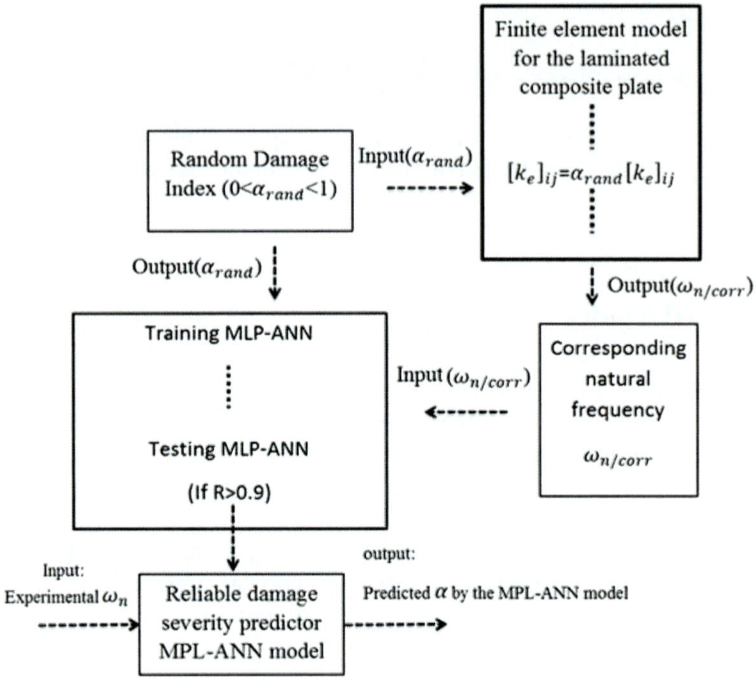

Fig. 4 Schematic of the proposed model of the current study

in this study. Each node has five degrees of freedom. The constant properties of the considered beams are: Length = width = 0.2, Height = 0.02, ρ = 1389 kg/m^3, E_2 = 9.65GPa, E_1 = 40E_2, G_{13} = 0.6 E_2, G_{23} = 0.5 E_2,v_{12} = 0.25 and $\upsilon_{12} = E_2/E_1\upsilon_{12}$.

As shown in Fig. 5, an element located at the mid of the plate with coordinate (i,j) (11,11) is damaged up to 70%.

Fig. 5 The considered single-damaged composite plate

4 Results

4.1 Effects of Single-Damage on Modal Characteristics

Table 1 shows results of five first natural frequencies for intact ($\alpha = 1$) damaged ($\alpha = 0.3$) rectangular laminated composite plates.

As seen in Table 1, the values of natural frequencies in the damaged rectangular composite plate are less than those in the intact one in all modes. These results are consistent with the results reported in the literature (see [15, 16]) and indicate our results' accuracy.

Figures 6, 7, 8, 9, 10 show the first five-mode shapes of intact and damaged laminated composite plates. As clearly seen, the single-damage causes a shift in

Table 1 Five first natural frequencies of intact ($\alpha = 1$) damaged ($\alpha = 0.3$) rectangular laminated composite plate

Frequency (Hz)	ω_1	ω_2	ω_3	ω_4	ω_5
$\alpha = 1$	1.54819	2.33739	2.99718	3.31531	3.77202
$\alpha = 0.3$	1.04571	1.60977	1.89952	2.34282	2.53130

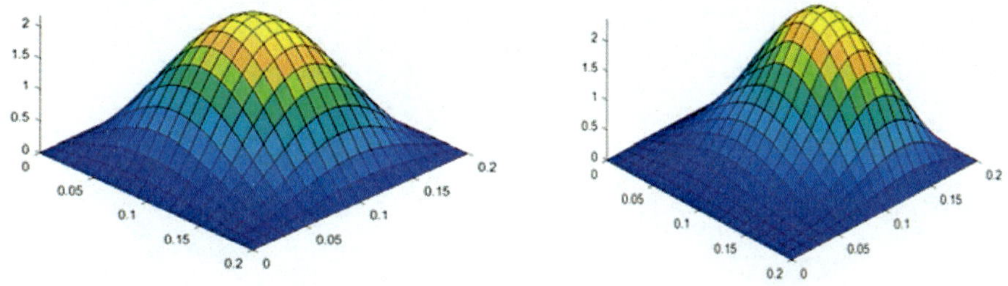

Fig. 6 The first mode shapes: Left: intact laminated composite plate ($\omega_1 = 1.54819$ Hz), Right: single-damaged laminated composite plate ($\omega_1 = 1.04571$ Hz)

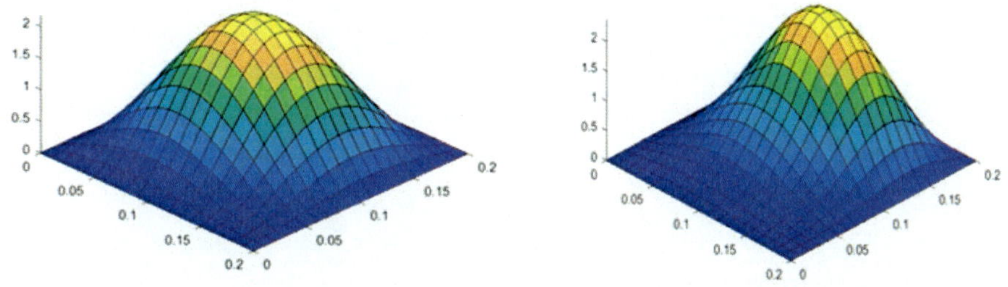

Fig. 7 The second mode shapes: Left: intact laminated composite plate ($\omega_2 = 2.33739$ Hz), Right: single-damaged laminated composite plate ($\omega_2 = 1.60977$ Hz)

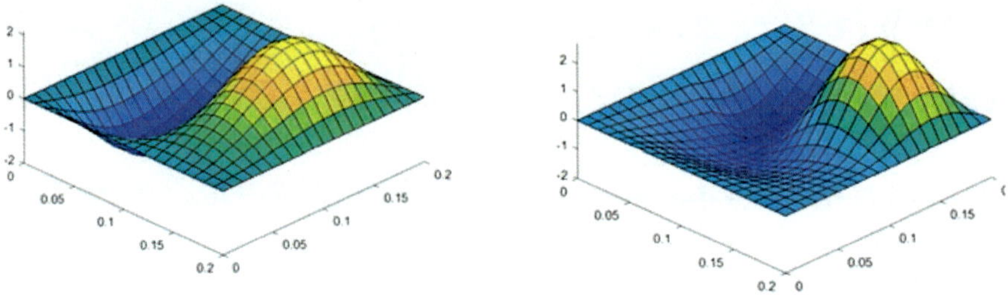

Fig. 8 The third mode shapes: Left: intact laminated composite plate ($\omega_3 = 2.99718$ Hz), Right: single-damaged laminated composite plate ($\omega_3 = 1.89952$ Hz)

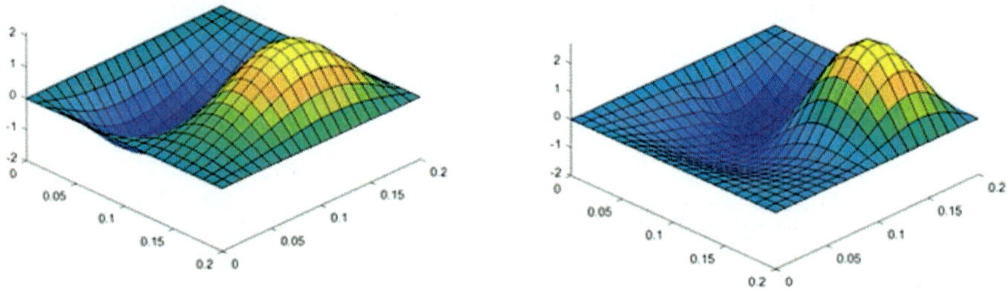

Fig. 9 The fourth mode shapes: Left: intact laminated composite plate ($\omega_4 = 3.31531$ Hz), Right: single-damaged laminated composite plate ($\omega_4 = 2.34282$ Hz)

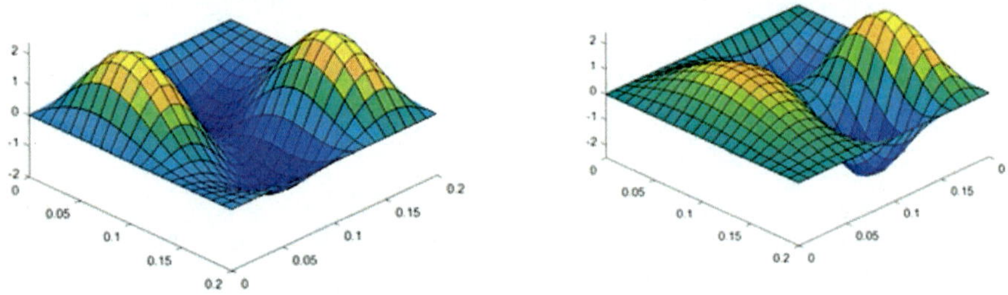

Fig. 10 The fifth mode shapes: Left: intact laminated composite plate ($\omega_5 = 3.77202$ Hz), Right: single-damaged laminated composite plate ($\omega_5 = 2.53130$ Hz)

mode shapes and increases instability. These findings are also consistent with the results reported in the literature (see [15, 16]) and indicate our results' accuracy.

4.2 *Performance of MPL-ANN*

In this part, the present paper's findings are presented based on data matching among inputs and outputs in testing and training data and the regression factor for our single damages scenario. Two hundred data as input (α_{rand}) are and 200 data as output ($\omega_{n/corr}$) are produced using FEM and as network's outputs and inputs.

Among these, 120 data are selected to train the ANN-MPL network and 80 data are selected to test the ANN-MPL network.

Figure 11 shows the real training data (red) vs. training data predicted by MLP ANN (blue). According to the figure, it is obvious that the real training data and the predicted training data by the MLP ANN are highly fit. Figure 12 indicates the actual test data (red) vs. test data predicted by the MLP ANN (blue). Based on the figure, it is obvious that the real test data and the predicted test data by the MLP ANN are highly fit.

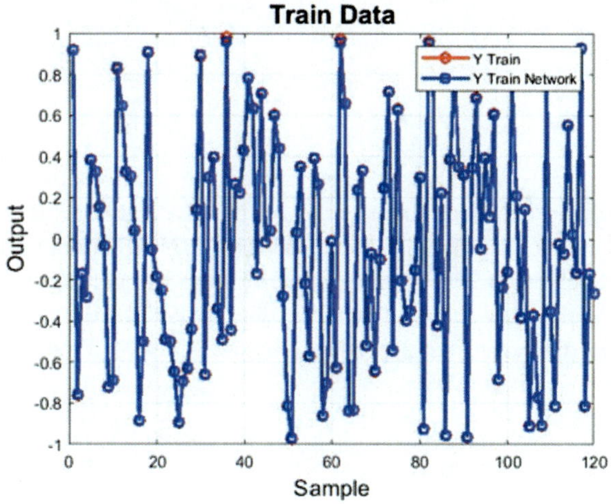

Fig. 11 Real training data (red) and training data predicted by MLP network (blue)

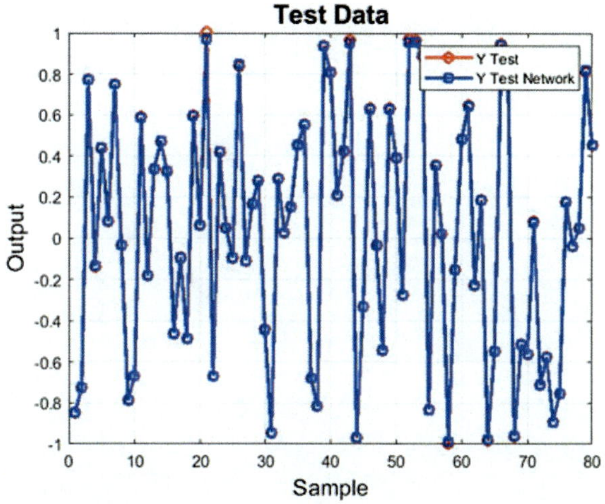

Fig. 12 Real testing data (red) vs. test data predicted by MLP network (blue)

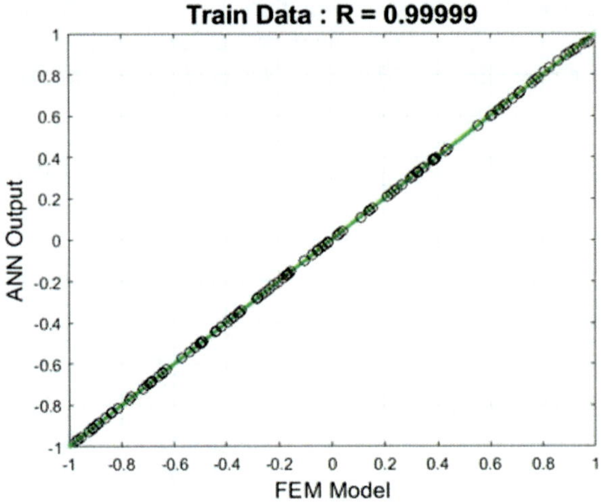

Fig. 13 Linear regression relationship among real training data and trained data predicted by MLP network

Figure 13 shows a linear regression relationship between real training data and trained data predicted by the ANN-MLP network. It indicates that the real training data vs. train data predicted by the ANN-MLP network are highly consistent since R-value is 0.99999. Also, Fig. 14 shows a linear regression relationship between real test data and the predicted test data by ANN-MLP network. It indicates that the real test data vs. test data predicted by the ANN-MLP network are highly consistent since R-value is 0.99997.

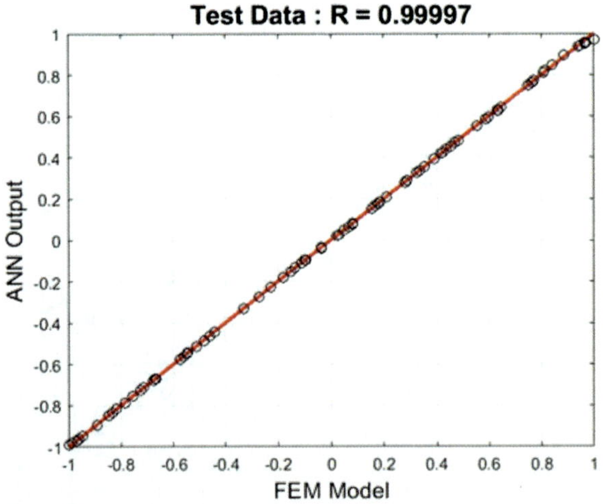

Fig. 14 Linear regression representation relationship among real test data and test data predicted by MLP network

5 Conclusions

In this paper, a novel method is proposed to detect severities of single damages in rectangular laminated composite plates using a multilayer perceptron (MLP) and the finite element method (FEM). A nine-node higher-order rectangular element with 5-DoFs is used to write the motion-based governing equation based on FEM. The FEM is applied to compare modal characteristics of the intact and single-damaged rectangular laminated composite plate. Results indicate that the natural frequency of single-damaged rectangular laminated composite plate is less than the intact rectangular laminated composite plate. These results are consistent with the results reported in the literature and indicate our results' accuracy. Also, the current paper's findings show that the single-damage causes a shift in mode shapes and increases instability. Also, the new model proposed in this study shows that the real training data vs. train data predicted by the ANN-MLP network are highly consistent since R-value is 0.99999. Our findings also indicate that the real test data vs. test data predicted by the ANN-MLP network are highly consistent since R-value is 0.99997. These findings demonstrate that the ANN-MLP network trained and tested in this paper is a reliable model to predict single-damage severity in the rectangular laminated composite plate by having experimental natural frequencies.

References

1. Khatir S, Tiachacht S, Le Thanh C, Khatir T, Capozucca R, Wahab MA (2020) Damage detection in laminated composite plates based on local frequency change ratio indicator. In: Proceedings of the 13th international conference on damage assessment of structures. Springer, Singapore, pp 887–898
2. Abedi M, Jafari-Talookolaei RA, Valvo PS (2016) A new solution method for free vibration analysis of rectangular laminated composite plates with general stacking sequences and edge restraints. Comput Struct 175:144–156
3. Jafari-Talookolaei RA (2015) Analytical solution for the free vibration characteristics of the rotating composite beams with a delamination. Aerosp Sci Technol 45:346–358
4. Yang JY, Xia BH, Chen Z, Li TL, Liu R (2020) Vibration-based structural damage identification: a review. Int J Robot Autom 35(2)
5. Maity D, Tripathy RR (2005) Damage assessment of structures from changes in natural frequencies using genetic algorithm. Struct Eng Mech 19(1):21–42
6. Dinh-Cong D, Vo-Duy T, Ho-Huu V, Nguyen-Thoi T (2019) Damage assessment in plate-like structures using a two-stage method based on modal strain energy change and Jaya algorithm. Inverse Probl Sci Eng 27(2):166–189
7. Gomes GF, da Cunha SS, Ancelotti AC (2019) A sunflower optimization (SFO) algorithm applied to damage identification on laminated composite plates. Eng Comp 35(2):619–626
8. Barman SK, Maiti DK, Maity D (2020) Damage detection of truss employing swarm-based optimization techniques: a comparison. In: Advanced engineering optimization through intelligent techniques. Springer, Singapore, pp 21–37
9. Gerist S, Maheri MR (2019) Structural damage detection using imperialist competitive algorithm and damage function. Appl Soft Comput 77:1–23
10. Liang RY, Hu J, Choy F (1992) Theoretical study of crack-induced eigenfrequency changes on beam structures. J Eng Mech 118(2):384–396

11. Khatir S, Tiachacht S, Thanh CL et al (2019) Damage assessment in composite laminates using ANN-PSO-IGA and Cornwell indicator. Compos Struct 230:111509
12. Reddy JN (2003) Mechanics of laminated composite plates and shells: theory and analysis. CRC Press, Boca Raton
13. Rhim J, Lee SW (1995) A neural network approach for damage detection and identification of structures. Comput Mech 16(6):437–443
14. Zainal-Mokhtar K, Mohamad-Saleh J (2013) An oil fraction neural sensor developed using electrical capacitance tomography sensor data. Sensors 13(9):11385–11406
15. Liang RY, Choy FK, Hu J (1991) Detection of cracks in beam structures using measurements of natural frequencies. J Franklin Inst 328(4):505–518
16. Wong CN, Zhu WD, Xu GY (2002) On an iterative general-order perturbation method for multiple structural damage detection. In: ASME International mechanical engineering congress and exposition. vol 36258, pp 301–313
17. Khatir S, Boutchicha D, Le Thanh C, Tran-Ngoc H, Nguyen TN, Abdel-Wahab M (2020) Improved ANN technique combined with Jaya algorithm for crack identification in plates using XIGA and experimental analysis. Theor Appl Fract Mech 102554
18. Fan W, Qiao P (2011) Vibration-based damage identification methods: a review and comparative study. Struct Health Monit 10(1):83–111
19. Tran-Ngoc H, Khatir S, De Roeck G, Bui-Tien T, Wahab MA (2019) An effi-cient artificial neural network for damage detection in bridges and beam-like structures by improving training parameters using cuckoo search algorithm. Eng Struct 199:109637

Adaptive Network-Based Fuzzy Inference for Damage Detection in Rectangular Laminated Composite Plates Using Vibrational Data

Morteza Saadatmorad, Ramazan-Ali Jafari-Talookolaei, Mohammad-Hadi Pashaei, Samir Khatir, and Magd Abdel Wahab

Abstract A novel model is developed to identify damages severities in rectangular laminated composite plates (RLCPs) based on the finite element method (FEM) and adaptive neuro-fuzzy inference system (ANFIS). Empirical data are generated via FEM, and then the ANFIS algorithm is applied to identify damages' severities of single damage scenarios. In this study, the networks' inputs and outputs are the first five natural frequencies and damages' severities. Also, a multilayer perceptron neural network (MLP) is developed to compare results obtained from ANFIS. Results show that the MLP model can identify the severity of every single damage scenario in RLCPs with a high precision ($R > 0.9$). Also, results obtained indicate that ANFIS can identify and predict the severity of every single damage scenario in RLCPs better than MLP with perfect precision ($R = 0.99999$).

Keywords Damaged laminated composites · Neuro-fuzzy inference system · Multilayer perceptron · Damaged structures · Finite element analysis

1 Introduction

In recent decades, composite elements have been broadly applied to build various structures because of their lightness, stiffness, and high strength [1]. Composite elements are also employed in various fields such as the aerospace engineering automotive industry [2]. Therefore, these materials are subjected to serve damages and defects that remarkably vary their structural performance [3]. Consequently, damage identification has drawn researchers' attention to propose useful and novel methodologies for monitoring various structures' health.

Usually, the research on damaged and imperfect structures is classified based on two main approaches: forward and inverse. In the forward approach, the dynamic

M. Saadatmorad · R.-A. Jafari-Talookolaei (✉) · M.-H. Pashaei
Babol Noshirvani University of Technology, Mazandaran province, Babol, Iran
e-mail: ra.jafari@nit.ac.ir

S. Khatir · M. Abdel Wahab
Soete Laboratory, Faculty of Engineering and Architecture, Ghent University, Ghent, Belgium

M. Abdel Wahab (ed.), *Proceedings of the 2nd International Conference on Structural Damage Modelling and Assessment*, Lecture Notes in Civil Engineering 204,
https://doi.org/10.1007/978-981-16-7216-3_14

properties such as natural frequencies and mode shapes are captured from vibrational structures. Hence, the location and severity of damages are predefined in this approach [4]. Also, the inverse approach is proposed to achieve the severity and position of damages by knowing the dynamic properties of vibrational structures [5]. In recent few decades, various strategies have been suggested as the inverse approach for vibrational structures. An example of the usual up-to-date strategies is the inverse approach based on artificial intelligence. These strategies employ soft computing algorithms like evolutional and data mining algorithms.

In the inverse approach based on artificial intelligence, Dinh-Cong et al. implemented the Jaya algorithm for identifying damaged plates [6]. Gomes et al. utilized the sunflower optimization method to detect damages of composite plate structures [7]. Khatir et al. [1] investigated damage detection in orthogonal composite plate structures. A novel damage index based on the local frequency change was proposed to evaluate damages within RLCPs with three layers. Their results revealed that the index could identify damage positions with great correctness. Numerous studies on damage identification based on optimization methods have been introduced in the recent few decades. Along with that, the ant colony algorithm was compared with the particle swarm algorithm by Barman et al. [8] to detect damages in plates and space trusses. Gerist et al. [9] used an algorithm called imperialist competitive (IC) for identifying damage through an innovative objective function based on mode shape.

Besides optimization-based schemes in the inverse approach, the methods based on artificial neural networks are commonly utilized in the literature. Artificial neural networks (ANNs) are advantageous to detecting damages in various structures. Liang et al. [10] employed normalized flexibility vectors as an ANN input to detect the damages' location and severity. Nasiri et al. employed ANN and modal analysis to detect the severity of a well-known type of inter-laminar damage termed delamination [11]. Yam et al. [12] introduced the application of neural networks and wavelet transform for identifying damages in composite structures. Khatir et al. [13] proposed a robust yet simple, novel method based on ANNs combined with an optimization method named Particle Swarm Optimization for detecting damage in composite plates through an index named Cornwell indicator. Yu et al. [14] introduced a novel online approach based on ANNs to identify composite shells' damages partly filled with fluid. The methods based adaptive network-based fuzzy inference (ANFIS) have recently been proposed as new methods in the literature's inverse approach. Nanda et al. [15] examined detecting multiple breathing cracks on fixed shafts via ANFIS and ANN. Findings showed that ANFIS identified multiple breathing cracks better than ANN. The rate of average error in ANFIS and ANN was 2.17% and 3.5%, respectively. Hakim et al. [16] presented an Adaptive neuro-fuzzy inference system (ANFIS) and artificial neural networks (ANNs) for structural damage identification. In this paper, the structure was a steel girder bridge using dynamic parameters. he proposed ANFIS architecture using hybrid learning algorithm was found to perform better than the multilayer feed-forward ANN which learns using the back-propagation algorithm.

Based on our literature review, ANFIS has been used in various strategies to detect damages in structures. However, this paper compares a multilayer perceptron neural network with a neuro-fuzzy inference system for the first time to predict the damages' severity in RLCPs by updating the finite element method.

2 Methodology

2.1 Mathematical Formulation

Consider the plates shown in Fig. 1. The figure shows two plates. one plate is intact (a) and the other is a damaged (b). Mathematical expression for the dynamic equation of an RLCP is obtained using the FEM. For this, an RLCP as a continuous system is discretized as finite elements. Local stiffness and mass matrices for each element are derived based on the first shear deformation theory (FSDT). Finally, global stiffness and mass matrices are obtained to express the equation of motion.

Constitutive Equations

According to the FSDT, displacement fields for each layer of the RLCP are presented as [17]:

$$\begin{aligned} u(x, y, z) &= u_0(x, y) + z\psi_x(x, y) \\ v(x, y, z) &= v_0(x, y) + z\psi_y(x, y) \\ w(x, y, z) &= w_0(x, y) \end{aligned} \tag{1}$$

where, u_i and $v_i (i = 0, 1, 2, 3)$ are the in-plane displacements of each layer of RLCP, as well as $w_j (j = 0, 1, 2)$ represent vertical displacements of each layer of the plate. ψ_x and ψ_y are rotations about x and y, respectively.

Fig. 1 Geometrical model of two rectangular laminated composite plates (RLCPs): **a** intact plate, **b** damaged plate

By having these displacements, generalized strains are obtained to express the constitutive equations as follows:

$$\begin{Bmatrix} N_x \\ N_y \\ N_{xy} \\ M_x \\ M_y \\ M_{xy} \end{Bmatrix} = \begin{bmatrix} A_{11} & A_{12} & A_{16} & B_{11} & B_{12} & B_{16} \\ A_{12} & A_{22} & A_{26} & B_{12} & B_{22} & B_{26} \\ A_{16} & A_{26} & A_{66} & B_{16} & B_{26} & B_{66} \\ B_{11} & B_{12} & B_{16} & D_{11} & D_{12} & D_{16} \\ B_{12} & B_{22} & B_{26} & D_{12} & D_{22} & D_{26} \\ B_{16} & B_{26} & B_{66} & D_{16} & D_{26} & D_{66} \end{bmatrix} \begin{Bmatrix} \varepsilon_x^0 \\ \varepsilon_y^0 \\ \gamma_{xy}^0 \\ \kappa_x \\ \kappa_y \\ \kappa_{xy} \end{Bmatrix} \tag{2}$$

$$\begin{Bmatrix} Q_{yz} \\ Q_{xz} \end{Bmatrix} = \begin{bmatrix} A_{44} & A_{45} \\ A_{45} & A_{55} \end{bmatrix} \begin{Bmatrix} \gamma_{yz} \\ \gamma_{xz} \end{Bmatrix} \tag{3}$$

where, A_{ij},B_{ij}, and $D_{ij}(i, j = 1, 2, \ldots, 6)$ are the extension, bending–extension coupling, and bending stiffness's, respectively. $A_{ij}(i, j = 4, 5)$ is the transverse shear stiffness's. A_{ij}, B_{ij}, and D_{ij} can be given as [12]:

$$A_{ij} = \sum_{k=1}^{N} \left(\overline{Q}_{ij}\right)_k (z_k - z_{k-1}) \quad \text{i, j} = 1, 2, 6 \tag{4}$$

$$B_{ij} = \frac{1}{2} \sum_{k=1}^{N} \left(\overline{Q}_{ij}\right)_k \left(z_k^2 - z_{k-1}^2\right) \quad \text{i, j} = 1, 2, 6 \tag{5}$$

$$D_{ij} = \frac{1}{3} \sum_{k=1}^{N} \left(\overline{Q}_{ij}\right)_k \left(z_k^3 - z_{k-1}^3\right) \quad \text{i, j} = 1, 2, 6 \tag{6}$$

$$A_{ij} = k_{ij} \sum_{k=1}^{N} \left(\overline{Q}_{ij}\right)_k (z_k - z_{k-1}) \quad \text{i, j} = 4, 5 \tag{7}$$

where, $\overline{Q}_{ij}$ is the transformed stiffness coefficients of the laminated composite layers, N is the total number of layers in the RLCP, z_k indicates the distance of the surface of the kth layer to the origin. In (7), k_{ij} is the shear correction coefficient ($k_{ij} = 0.833$).

By the above assumptions, the considered RLCP is converted into one lamina that the Eqs. (2) and (3) are governed on it. As a result, the Eqs. (4)-(7) present the equivalent stiffness's.

Development of Finite Element Equations

In this paper, a nine-node higher-order rectangular element is considered, as shown in Fig. 2.

Each node has five degrees of freedom (u_0,v_0,w_0,ψ_x, andψ_y). Where the displacement field functionsu_0,v_0,w_0,ψ_x, and ψ_y are written in terms of nodal displacement

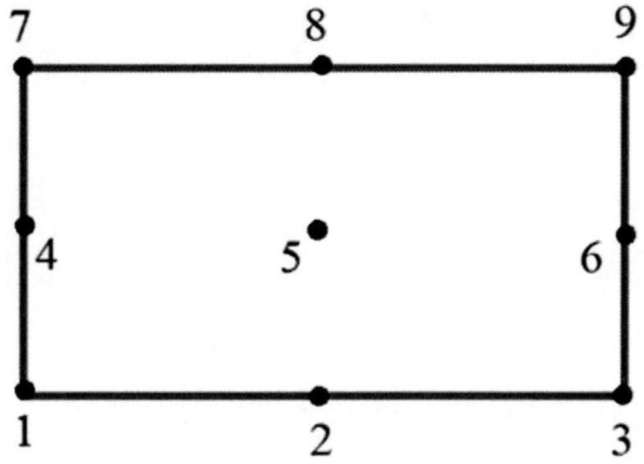

Fig. 2 The considered higher-order rectangular element with nine nodes

vector $\{d\}_{45\times 1}$ and the shape functionsN_{ψ_y}, N_{ψ_x}, N_w,N_v, andN_u. $\{d\}_{45\times 1}$ is presented as:$\{d\}_{45\times 1} = \{u_1, v_1, w_1, \psi_{x1}, \psi_{y1}, ..., u_9, v_9, w_9, \psi_{x9}, \psi_{y9}\}^T$.

The Lagrange interpolation is used to find the value of each node's degree of freedom. To obtain local stiffness and mass matrices, first, the potential and kinetic energies of the considered RLCP are obtained based on FSDT theory, then; rewrite them in terms of nodal displacements.

$$\begin{aligned}
U_P = 0.5\int_0^a\int_0^b\Big[& A_{11}u_{0,x}^2 + A_{22}v_{0,y}^2 + A_{66}\left(u_{0,y}^2 + v_{0,x}^2 + 2u_{0,y}v_{0,x}\right) + D_{11}\psi_{x,x}^2 \\
& + D_{22}\psi_{y,y}^2 + D_{66}\left(\psi_{x,y}^2 + \psi_{y,x}^2 + 2\psi_{x,y}\psi_{y,x}\right) + A_{44}\left(\psi_y^2 + w_{0,y}^2 + 2\psi_y w_{0,y}\right) \\
& + A_{55}\left(\psi_x^2 + w_{0,x}^2 + 2\psi_x w_{0,x}\right) + 2A_{12}u_{0,x}v_{0,y} + 2A_{16}\left(u_{0,y} + v_{0,x}\right)u_{0,x} + 2B_{11}\psi_{x,x}u_{0,x} \\
& + 2B_{12}\psi_{y,y}u_{0,x} + 2B_{16}\left(\psi_{x,y} + \psi_{y,x}\right)u_{0,x} + 2A_{26}\left(u_{0,y} + v_{0,x}\right)v_{0,y} + 2B_{12}\psi_{x,x}v_{0,y} \\
& + 2B_{22}\psi_{y,y}v_{0,y} + 2B_{26}\left(\psi_{x,y} + \psi_{y,x}\right)v_{0,y} + 2B_{16}\psi_{x,x}\left(u_{0,y} + v_{0,x}\right) \\
& + 2B_{26}\psi_{y,y}\left(u_{0,y} + v_{0,x}\right) + 2B_{66}\left(\psi_{x,y} + \psi_{y,x}\right)\left(u_{0,y} + v_{0,x}\right) + 2D_{12}\psi_{y,y}\psi_{x,x} \\
& + 2D_{16}\left(\psi_{x,y} + \psi_{y,x}\right)\emptyset_{,x} + 2D_{26}\left(\psi_{x,y} + \psi_{y,x}\right)\psi_{y,y} + 2A_{45}\left(\psi_x + w_{0,x}\right)\left(\psi_y + w_{0,y}\right)\Big]\text{dy dx}
\end{aligned} \tag{8}$$

It can be expressed in terms of local stiffness matrix $[k_e]$ as:

$$U_P = \frac{1}{2}\{d\}_{45\times 1}^T \cdot [k_e]_{45\times 45}\{d\}_{45\times 1} \tag{9}$$

As a result, the local stiffness matrix $[k_e]$ can be written as:

$$
\begin{aligned}
&[k_e] = \int_0^a \int_0^b [\ A_{11}\lfloor N_{u_{0,x}}\rfloor^T . \lfloor N_{u_{0,x}}\rfloor + A_{22}\lfloor N_{v_{0,y}}\rfloor^T . \lfloor N_{v_{0,y}}\rfloor + A_{66}(\lfloor N_{u_{0,y}}\rfloor^T . \lfloor N_{u_{0,y}}\rfloor + \\
&\lfloor N_{v_{0,x}}\rfloor^T . \lfloor N_{v_{0,x}}\rfloor + \lfloor N_{u_{0,y}}\rfloor^T . \lfloor N_{v_{0,x}}\rfloor + \lfloor N_{v_{0,x}}\rfloor^T . \lfloor N_{u_{0,y}}\rfloor A_{66}(\lfloor N_{u_{0,y}}\rfloor^T . \lfloor N_{u_{0,y}}\rfloor + \\
&\lfloor N_{v_{0,x}}\rfloor^T . \lfloor N_{v_{0,x}}\rfloor + \lfloor N_{u_{0,y}}\rfloor^T . \lfloor N_{v_{0,x}}\rfloor + \lfloor N_{v_{0,x}}\rfloor^T . \lfloor N_{u_{0,y}}\rfloor) + D_{11}\lfloor N_{\psi_{x,x}}\rfloor^T . \lfloor N_{\psi_{x,x}}\rfloor + \\
&D_{22}\lfloor N_{\psi_{y,y}}\rfloor^T . \lfloor N_{\psi_{y,y}}\rfloor + D_{66}\left(\lfloor N_{\psi_{x,y}}\rfloor^T . \lfloor N_{\psi_{x,y}}\rfloor + \lfloor N_{\psi_{y,x}}\rfloor^T . \lfloor N_{\psi_{y,x}}\rfloor + \right.\\
&\left.\lfloor N_{\psi_{xy}}\rfloor^T . \lfloor N_{\psi_{y,x}}\rfloor + \lfloor N_{\psi_{y,x}}\rfloor^T . \lfloor N_{\psi_{x,y}}\rfloor\right) + A_{44}\left(\lfloor N_{\psi_y}\rfloor^T . \lfloor N_{\psi_y}\rfloor + \lfloor N_{w_{0,y}}\rfloor^T . \lfloor N_{w_{0,y}}\rfloor + \right.\\
&\left.\lfloor N_{\psi_y}\rfloor^T . \lfloor N_{w_{0,y}}\rfloor + \lfloor N_{w_{0,y}}\rfloor^T . \lfloor N_{\psi_y}\rfloor\right) A_{44}\left(\lfloor N_{\psi_y}\rfloor^T . \lfloor N_{\psi_y}\rfloor + \lfloor N_{w_{0,y}}\rfloor^T . \lfloor N_{w_{0,y}}\rfloor + \right.\\
&\left.\lfloor N_{\psi_y}\rfloor^T . \lfloor N_{w_{0,y}}\rfloor + \lfloor N_{w_{0,y}}\rfloor^T . \lfloor N_{\psi_y}\rfloor\right) A_{44}\left(\lfloor N_{\psi_y}\rfloor^T . \lfloor N_{\psi_y}\rfloor + \lfloor N_{w_{0,y}}\rfloor^T . \lfloor N_{w_{0,y}}\rfloor + \right.\\
&\left.\lfloor N_{\psi_y}\rfloor^T . \lfloor N_{w_{0,y}}\rfloor + \lfloor N_{w_{0,y}}\rfloor^T . \lfloor N_{\psi_y}\rfloor\right) + A_{55}\left(\lfloor N_{\psi_x}\rfloor^T . \lfloor N_{\psi_x}\rfloor + \lfloor N_{w_{0,x}}\rfloor^T . \lfloor N_{w_{0,x}}\rfloor + \right.\\
&\left.\lfloor N_{\psi_x}\rfloor^T . \lfloor N_{w_{0,x}}\rfloor + \lfloor N_{w_{0,x}}\rfloor^T . \lfloor N_{\psi_x}\rfloor\right) + A_{12}(\lfloor N_{u_{0,x}}\rfloor^T . \lfloor N_{v_{0,y}}\rfloor + \lfloor N_{v_{0,y}}\rfloor^T . \lfloor N_{u_{0,x}}\rfloor + \\
&A_{16}(\lfloor N_{u_{0,y}}\rfloor^T . \lfloor N_{u_{0,x}}\rfloor + \lfloor N_{u_{0,x}}\rfloor^T . \lfloor N_{u_{0,y}}\rfloor + \lfloor N_{v_{0,x}}\rfloor^T . \lfloor N_{u_{0,x}}\rfloor + \lfloor N_{u_{0,x}}\rfloor^T . \lfloor N_{v_{0,x}}\rfloor + \\
&B_{11}\left(\lfloor N_{\psi_{x,x}}\rfloor^T . \lfloor N_{u_{0,x}}\rfloor + \lfloor N_{u_{0,x}}\rfloor^T . \lfloor N_{\psi_{x,x}}\rfloor\right) + B_{12}\left(\lfloor N_{\psi_{y,y}}\rfloor^T . \lfloor N_{u_{0,x}}\rfloor + \right.\\
&\left.\lfloor N_{u_{0,x}}\rfloor^T . \lfloor N_{\psi_{y,y}}\rfloor\right) B_{12}\left(\lfloor N_{\psi_{y,y}}\rfloor^T . \lfloor N_{u_{0,x}}\rfloor + \lfloor N_{u_{0,x}}\rfloor^T . \lfloor N_{\psi_{y,y}}\rfloor\right) + \\
&B_{16}(\lfloor N_{\psi_{x,y}}\rfloor^T . \lfloor N_{u_{0,x}}\rfloor + \lfloor N_{u_{0,x}}\rfloor^T . \lfloor N_{\psi_{x,y}}\rfloor + \lfloor N_{\psi_{y,x}}\rfloor^T . \lfloor N_{u_{0,x}}\rfloor + \lfloor N_{u_{0,x}}\rfloor^T . \lfloor N_{\psi_{y,x}}\rfloor) + \\
&A_{26}(\lfloor N_{u_{0,y}}\rfloor^T . \lfloor N_{v_{0,y}}\rfloor + \lfloor N_{v_{0,y}}\rfloor^T . \lfloor N_{u_{0,y}}\rfloor + \lfloor N_{v_{0,x}}\rfloor^T . \lfloor N_{v_{0,y}}\rfloor + \\
&\lfloor N_{v_{0,y}}\rfloor^T . \lfloor N_{v_{0,x}}\rfloor) + B_{12}\left(\lfloor N_{\psi_{x,x}}\rfloor^T . \lfloor N_{v_{0,y}}\rfloor + \lfloor N_{v_{0,y}}\rfloor^T . \lfloor N_{\psi_{x,x}}\rfloor\right) + \\
&B_{26}\left(\lfloor N_{\psi_{x,y}}\rfloor^T . \lfloor N_{v_{0,y}}\rfloor + \lfloor N_{v_{0,y}}\rfloor^T . \lfloor N_{\psi_{x,y}}\rfloor + \lfloor N_{\psi_{y,x}}\rfloor^T . \lfloor N_{v_{0,y}}\rfloor + \lfloor N_{v_{0,y}}\rfloor^T . \lfloor N_{\psi_{y,x}}\rfloor\right) + \\
&B_{26}\left(\lfloor N_{\psi_{x,y}}\rfloor^T . \lfloor N_{v_{0,y}}\rfloor + \lfloor N_{v_{0,y}}\rfloor^T . \lfloor N_{\psi_{x,y}}\rfloor + \lfloor N_{\psi_{y,x}}\rfloor^T . \lfloor N_{v_{0,y}}\rfloor + \lfloor N_{v_{0,y}}\rfloor^T . \lfloor N_{\psi_{y,x}}\rfloor\right) + \\
&+B_{16}(\lfloor N_{\psi_{x,x}}\rfloor^T . \lfloor N_{u_{0,y}}\rfloor + \lfloor N_{u_{0,y}}\rfloor^T . \lfloor N_{\psi_{x,x}}\rfloor + \lfloor N_{\psi_{x,x}}\rfloor^T + . \lfloor N_{v_{0,x}}\rfloor + \\
&\lfloor N_{v_{0,y}}\rfloor^T . \lfloor N_{\psi_{x,y}}\rfloor) + \\
&B_{26}(\lfloor N_{\psi_{y,y}}\rfloor^T . \lfloor N_{u_{0,y}}\rfloor + \lfloor N_{u_{0,y}}\rfloor^T . \lfloor N_{\psi_{y,y}}\rfloor + \lfloor N_{\psi_{y,y}}\rfloor^T . \lfloor N_{v_{0,x}}\rfloor + \lfloor N_{v_{0,x}}\rfloor^T . \lfloor N_{\psi_{y,y}}\rfloor) + \\
&B_{66}(\lfloor N_{\psi_{x,y}}\rfloor^T . \lfloor N_{u_{0,y}}\rfloor + \lfloor N_{u_{0,y}}\rfloor^T . \lfloor N_{\psi_{x,y}}\rfloor + \lfloor N_{\psi_{x,y}}\rfloor^T . \lfloor N_{v_{0,x}}\rfloor + \\
&\lfloor N_{v_{0,x}}\rfloor^T . \lfloor N_{\psi_{x,y}}\rfloor + \lfloor N_{\psi_{y,x}}\rfloor^T . \lfloor N_{u_{0,y}}\rfloor + \lfloor N_{u_{0,y}}\rfloor^T . \lfloor N_{\psi_{y,x}}\rfloor + \lfloor N_{\psi_{y,x}}\rfloor^T . \lfloor N_{v_{0,x}}\rfloor + \\
&\lfloor N_{v_{0,x}}\rfloor^T . \lfloor N_{\psi_{y,x}}\rfloor) + D_{26}(\lfloor N_{\psi_{x,y}}\rfloor^T . \lfloor N_{\psi_{y,y}}\rfloor + \lfloor N_{\psi_{y,y}}\rfloor^T . \lfloor N_{\psi_{x,y}}\rfloor + \lfloor N_{\psi_{y,x}}\rfloor^T . \lfloor N_{\psi_{y,y}}\rfloor) + \\
&A_{45}(\lfloor N_{\psi_x}\rfloor^T . \lfloor N_{\psi_y}\rfloor + \lfloor N_{\psi_y}\rfloor^T . \lfloor N_{\psi_x}\rfloor + \lfloor N_{\psi_x}\rfloor^T . \lfloor N_{w_{0,y}}\rfloor + \lfloor N_{w_{0,y}}\rfloor^T . \lfloor N_{\psi_x}\rfloor + \lfloor N_{w_{0,x}}\rfloor^T . \lfloor N_{\psi_y}\rfloor \\
&+ \lfloor N_{\psi_y}\rfloor^T . \lfloor N_{w_{0,x}}\rfloor + \lfloor N_{w_{0,x}}\rfloor^T . \lfloor N_{w_{0,y}}\rfloor + \lfloor N_{w_{0,y}}\rfloor^T . \lfloor N_{w_{0,x}}\rfloor)]\,\mathrm{dydx}
\end{aligned}
\tag{10}
$$

The kinetic energy of in one considered element of the RLCP based on the FSDT is expressed as follows:

$$\mathrm{T} = \frac{1}{2}\int_0^a \int_0^b \left[I_0\left(u_{,t}^2 + v_{,t}^2 + w_{,t}^2\right) + 2I_1\left(u_{,t}\psi_{x,t} + v_{,t}\psi_{y,t}\right) \right. \\ \left. + I_2\left(\psi_{x,t}^2 + \psi_{y,t}^2\right)\right] dy dx \tag{11}$$

where (I_0, I_1, I_2) are the mass moment of inertia:

$$(I_0, I_1, I_2) = \int_{-h/2}^{h/2} \rho(z)\left(1, z, z^2\right) dz \tag{12}$$

It can be expressed in terms of local mass matrix $[m_e]$ as:

$$\mathrm{T} = \frac{1}{2}\{\dot{d}\}^T \cdot [m_e] \cdot \{\dot{d}\} \tag{13}$$

where

$$[M_e] = \int_0^a \int_0^b \left[I_0\left(\lfloor N_u \rfloor^T \cdot \lfloor N_u \rfloor + \lfloor N_v \rfloor^T \cdot \lfloor N_v \rfloor + \lfloor N_w \rfloor^T \cdot \lfloor N_w \rfloor\right) \right. \\ + I_1\left(\lfloor N_u \rfloor^T \cdot \lfloor N_{\psi_x} \rfloor + \lfloor N_{\psi_x} \rfloor^T \cdot \lfloor N_u \rfloor + \lfloor N_v \rfloor^T \cdot \lfloor N_{\psi_y} \rfloor + \lfloor N_{\psi_y} \rfloor^T \cdot \lfloor N_v \rfloor\right) \\ \left. + I_2\left(\lfloor N_{\psi_x} \rfloor^T \cdot \lfloor N_{\psi_x} \rfloor + \lfloor N_{\psi_y} \rfloor^T \cdot \lfloor N_{\psi_y} \rfloor\right)\right] dy dx$$

Governing Equation of Motion

Global stiffness matrix [K] and global mass matrix $[M]$ are obtained by assembling local stiffness matrices [$\mathrm{k_e}$] of all elements and local mass matrices [$\mathrm{m_e}$] of all elements, respectively. Finally, the governing equation of motion based on FEM is expressed as follows:

$$[\mathrm{M}]\{\ddot{\Delta}\} + [\mathrm{K}]\{\Delta\} = 0 \tag{14}$$

where $\{\Delta\}$ includes the degrees of freedom for all nodes in system. By assuming $\{\Delta\}=\{\Delta_0\}e^{i\omega t}$ as well as $\lambda = \omega^2$, it can be written:

$$([\mathrm{K}] - \lambda[\mathrm{M}])\{\Delta_0\} = 0 \tag{15}$$

where ω denotes the natural frequency and $\{\Delta_0\}$ is the corresponding modes shapes to the natural frequencies. The non-zero solution of the system is calculated through the following equation using Matlab software.

$$\det([K] - \lambda[M]) = 0 \tag{16}$$

2.2 *Adaptive Network-Based Fuzzy Inference*

Adaptive network-based fuzzy inference system (ANFIS) is a combination of neural networks and the fuzzy inference system. It is an approach that uses a neural network for optimizing fuzzy inference systems. To generate outputs corresponding to inputs, ANFIS uses if–then rules and proper membership functions. First, primary membership functions and fuzzy rules are set based on human expertise concerning the actual outputs. Then, ANFIS can adjust these fuzzy membership functions and if–then rules by minimizing output error.

The approach of ANFIS is applied as a beneficial tool for solving problems of function approximation. A typical architecture of ANFIS is depicted in Fig. 3, which involves four rules, two inputs, and one output. It is the first-order model of the Sugeno fuzzy system. For convenience, in this mode, it is supposed that each input has two membership functions.

In Fig. 3, A_i and B_j (i,j = 1,2) denote symbols of membership functions related to the inputs I_{n1} and I_{n2}, respectively. Also, w_{ij} and $\overline{w}_{ij}$ represent weight factors.

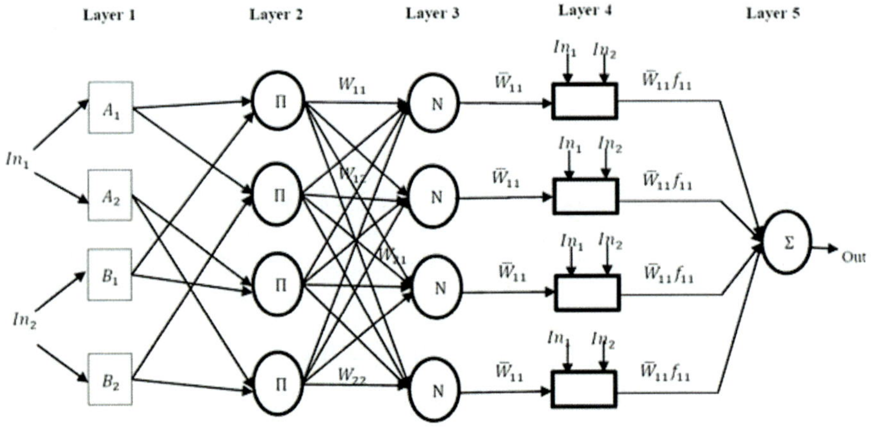

Fig. 3 A typical first-order model of the Sugeno fuzzy system [18]

3 Proposed Model

This work proposes a new algorithm to identify the severity of the single damages of RLCPs using FEM and ANFIS. First, the finite element governing equation is formed for the intact RLCPs to calculate natural frequencies and mode shapes.

Then, single damage is applied to the RLCP to obtain five first natural frequencies and their corresponding mode shapes in a single-damage scenario and investigate the effect of single-damage in the RLCP.

As seen in Fig. 4, a damage index called α_{rand} is introduced as the coefficient of the local stiffness matrix $[k_e]_{i,j}$ to represent the presence of a single-damage in the location of (i,j) in the RLCP. The value of the damage index is between 0 to 1 (i.e., $0 < \alpha_{rand} < 1$) because if = 0, then there is no local element in the location (i,j), and if = 1, then there is no damage in the local element in the location of (i,j). 100 α_{rand} s are produces randomly.

Two hundred α_{rand} are produced randomly to be used as the coefficients of $[k_e]_{i,j}$; thus, two hundred corresponding natural frequencies ($\omega_{n/corr}$) are produced. Therefore, α_{rand} and $\omega_{n/corr}$ are considered as the inputs and outputs for our finite element model, respectively. Then, as shown in Fig. 4, α_{rand} and $\omega_{n/corr}$ are used as the outputs and inputs for our ANFIS, respectively, to train and test its performance. When the regression factor(R) is more than 0.9, our ANFIS will be a high-performance network. Therefore, it can predict every single-damage severity by having experimental natural frequencies of the RLCP.

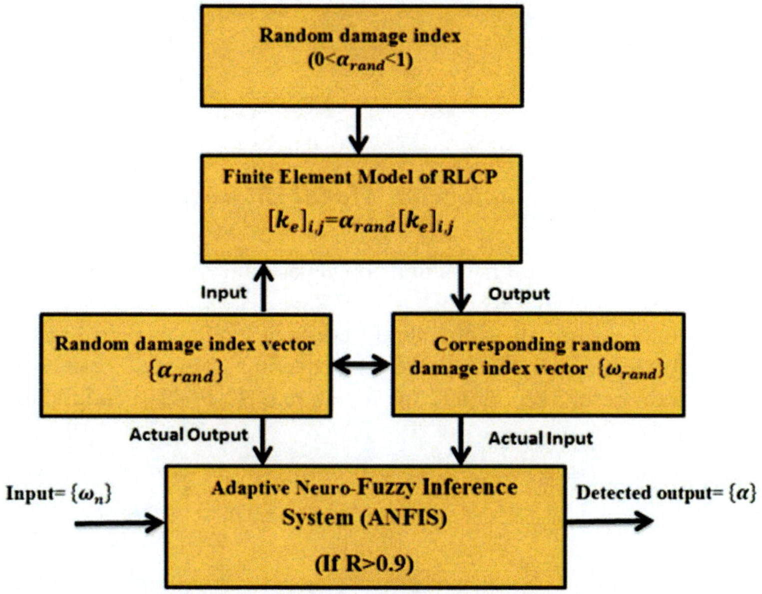

Fig. 4 Schematic of the proposed procedure of the current work

Fig. 5 The considered single-damaged composite plate

3.1 Computational Examples

A two-layer RLCP $[0/\frac{\pi}{4}]$ with boundary conditions clamp-clamp-clamp-clamp (CCCC) is discretized into 121 finite elements to investigate the effects of single-damage on the RLCP's modal characteristics. As mentioned, a nine-node higher-order rectangular element is used in this study. Each node has five degrees of freedom. The constant properties of the considered plates are: Length = width = 0.2, Height = 0.02, $\rho = 1389\mathrm{kg/m^3}$, E_2=9.65GPa, $E_1 = 40E_2$, G_{13}=0.6 E_2, G_{23}=0.5 E_2,$\nu_{12} = 0.25$, and $\nu_{12} = E_2/E_1\nu_{12}$.

As shown in Fig. 5, an element located at the mid of the plate with coordinate (i,j) (11,11) is damaged up to 70%.

4 Results

4.1 Effects of Single-Damage on Modal Characteristics

Table 1 shows results of five first natural frequencies for intact ($\alpha = 1$) damaged ($\alpha = 0.3$) RLCPs.

As seen in Table 1, the values of natural frequencies in the damaged rectangular composite plate are less than those in the intact one in all modes. These results are consistent with the results reported in the literature (see [17–20]) and indicate our results' accuracy.

Table 1 Five first natural frequencies of intact ($\alpha = 1$) damaged ($\alpha = 0.3$) RLCPs

Frequency (Hz)	ω_1	ω_2	ω_3	ω_4	ω_5
$\alpha = 1$	1.54819	2.33739	2.99718	3.31531	3.77202
$\alpha = 0.3$	1.04571	1.60977	1.89952	2.34282	2.53130

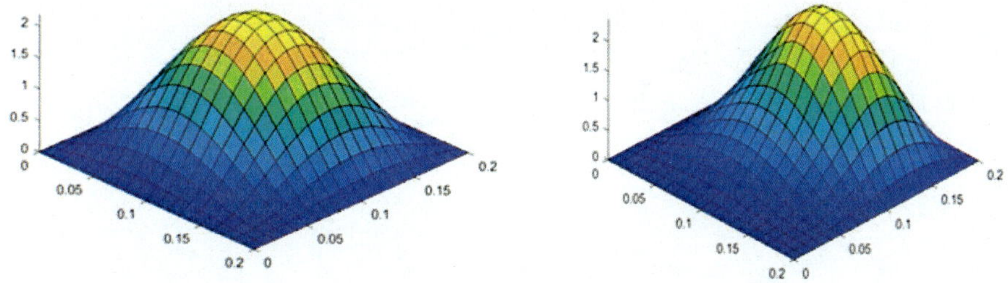

Fig. 6 The first mode shapes: Left: intact RLCP ($\omega_1 = 1.54819$Hz), Right: single-damaged RLCP ($\omega_1 = 1.04571$Hz)

Figures 6, 7, 8, 9, and 10 show the first five-mode shapes of intact and damaged RLCP. As clearly seen, the single-damage causes a shift in mode shapes and increases instability. These findings are also consistent with the results reported in the literature (see [18–20]) and indicate our results' accuracy.

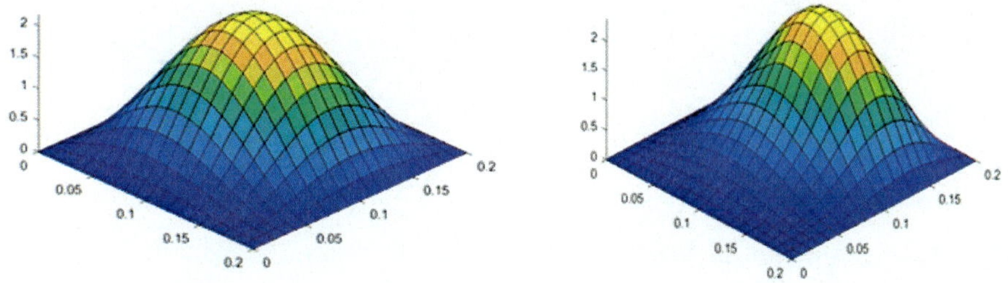

Fig. 7 The second mode shapes: Left: intact RLCP ($\omega_2 = 2.33739$Hz), Right: single-damaged RLCP ($\omega_2 = 1.60977$Hz)

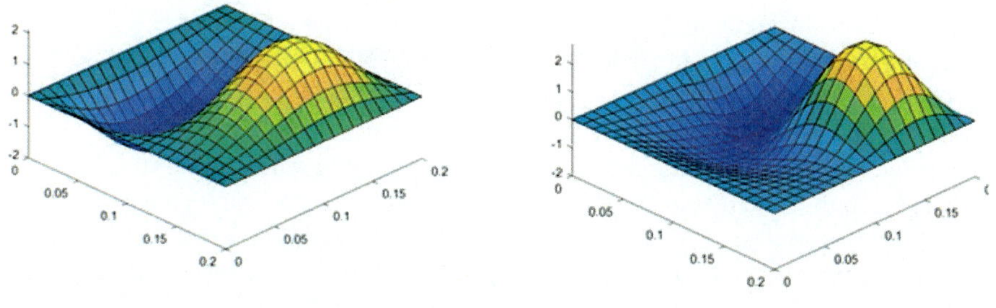

Fig. 8 The third mode shapes: Left: intact RLCP (ω_3=2.99718 Hz), Right: single-damaged RLCP (ω_3=1.89952 Hz)

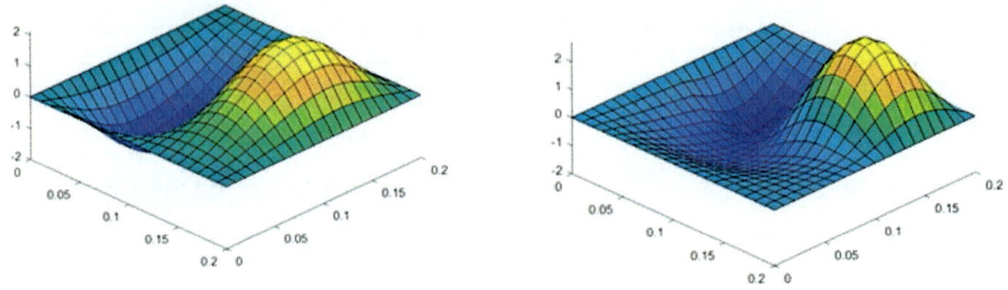

Fig. 9 The fourth mode shapes: Left: intact RLCP (ω_4=3.31531 Hz), Right: single-damaged RLCP (ω_4=2.34282 Hz)

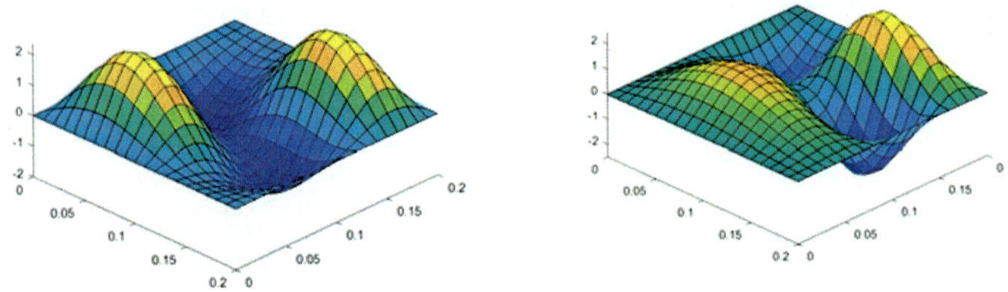

Fig. 10 The fifth mode shapes: Left: intact RLCP (ω_5=3.77202 Hz), Right: single-damaged RLCP (ω_5=2.53130 Hz)

4.2 Performance of MPL-ANN

In this part, the present paper's findings are presented based on data matching among inputs and outputs in testing and training data and the regression factor for our single damages scenario. Two hundred data as input (α_{rand}) are and 200 data as output ($\omega_{n/corr}$) are produced using FEM and as network's outputs and inputs.

Among these, 120 data are selected to train the ANN-MPL network and 80 data are selected to test the ANN-MPL network.

Figure 11 shows the real training data (red) versus training data predicted by MLP ANN (blue). According to the figure, it is obvious that the real training data and the predicted training data by the MLP ANN are highly fit. Figure 12 indicates the actual test data (red) versus test data predicted by the MLP ANN (blue). Based on the figure, it is obvious that the real test data and the predicted test data by the MLP ANN are highly fit.

Figure 13 shows a linear regression relationship between real training data and trained data predicted by the ANN-MLP network. It indicates that the real training data versus train data predicted by the ANN-MLP network are highly consistent since R-value is 0.99999. Also, Fig. 14 shows a linear regression relationship between real test data and the predicted test data by ANN-MLP network. It indicates that the real

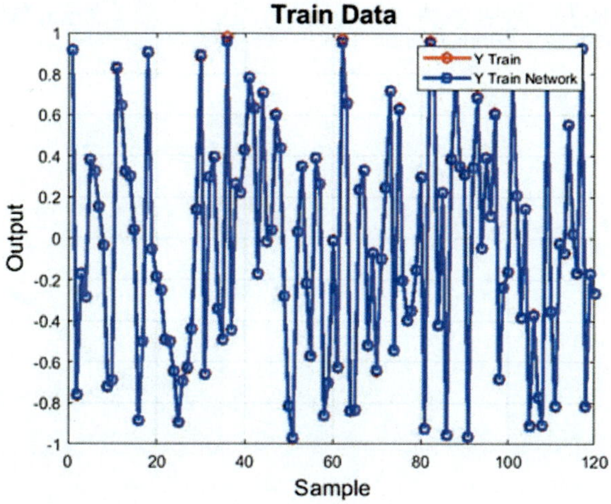

Fig. 11 Real training data (red) and training data predicted by MLP network (blue)

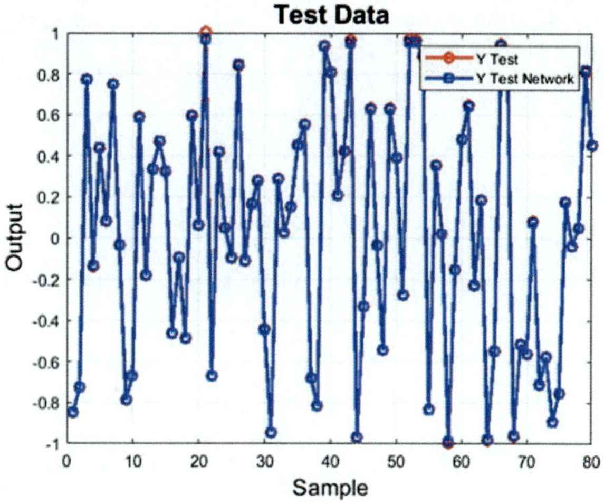

Fig. 12 Real testing data (red) versus test data predicted by MLP network (blue)

test data versus test data predicted by the ANN-MLP network are highly consistent since R-value is 0.99997.

4.3 Performance of ANFIS

As mentioned, in this work, the current paper's results are presented based on data matching among inputs and outputs in testing and training data and the regression factor for our single damages scenario. Two hundred data as input (α_{rand}) are and

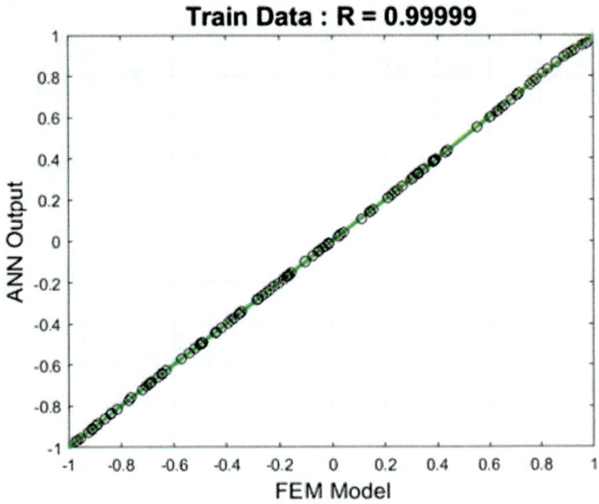

Fig. 13 Linear regression relationship among real training data and trained data predicted by MLP network

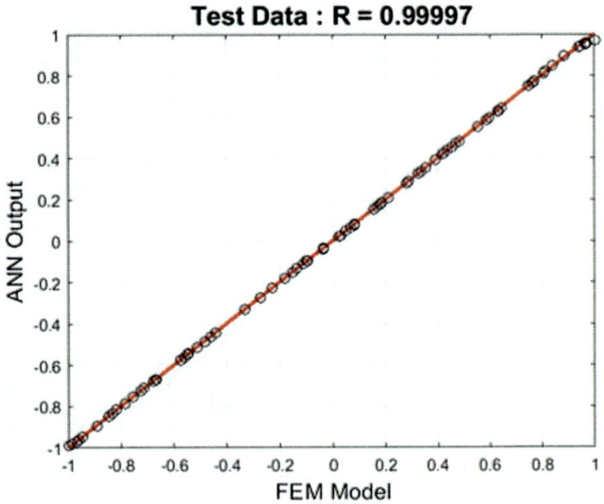

Fig. 14 Linear regression representation relationship among real test data and test data predicted by MLP network

200 data as output ($\omega_{n/corr}$) are produced using FEM and as network's outputs and inputs. In this section, performance of ANFIS is evaluated.

Among these, 120 data are selected to train the ANFIS and 80 data are selected to test the ANFIS.

Figure 15 shows the real training data (red) versus training data predicted by ANFIS (blue). According to the figure, it is obvious that the real training data and the predicted training data by the ANFIS are highly fit. Figure 16 indicates the actual test data (red) versus test data predicted by the ANFIS (blue). Based on the figure, it is obvious that the real test data and the predicted test data by the ANFIS are highly fit.

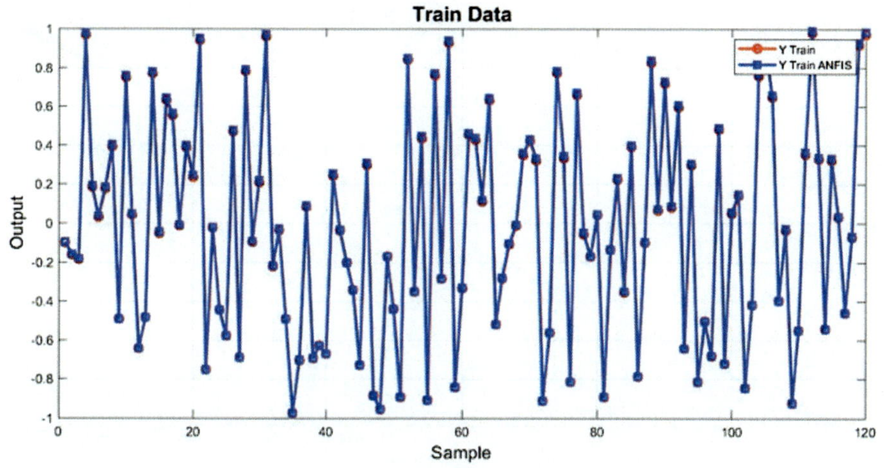

Fig. 15 Real training data (red) and training data predicted by ANFIS (blue)

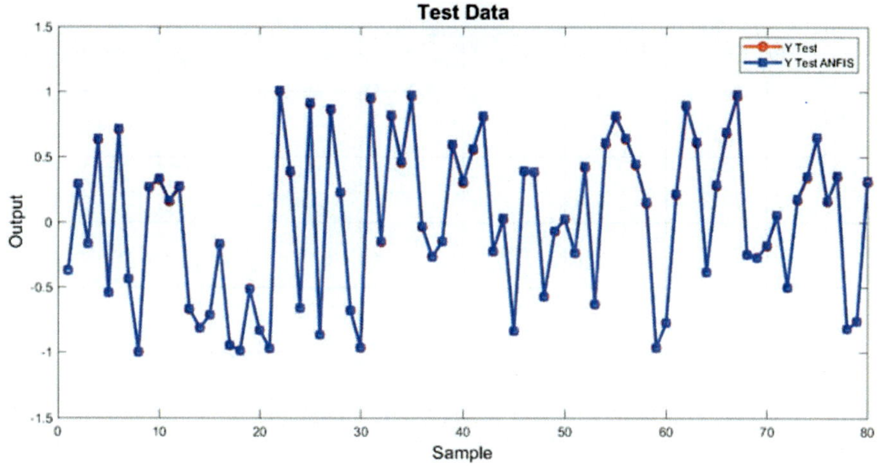

Fig. 16 Real testing data (red) versus test data predicted by ANFIS (blue)

Figure 17 shows a linear regression relationship between real training data and trained data predicted by the ANFIS. It indicates that the real training data versus train data predicted by the ANFIS network are perfect consistent since R-value is approximately 1. Also, Fig. 18 shows a linear regression relationship between real test data and the predicted test data by ANFIS. It indicates that the real test data versus test data predicted by the ANFIS are perfect consistent since R-value is 1.

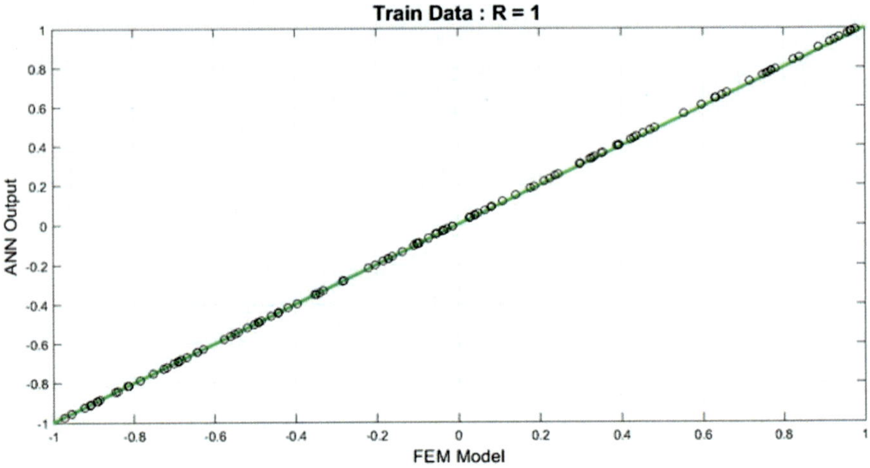

Fig. 17 Linear regression relationship among real training data and trained data predicted by ANFIS

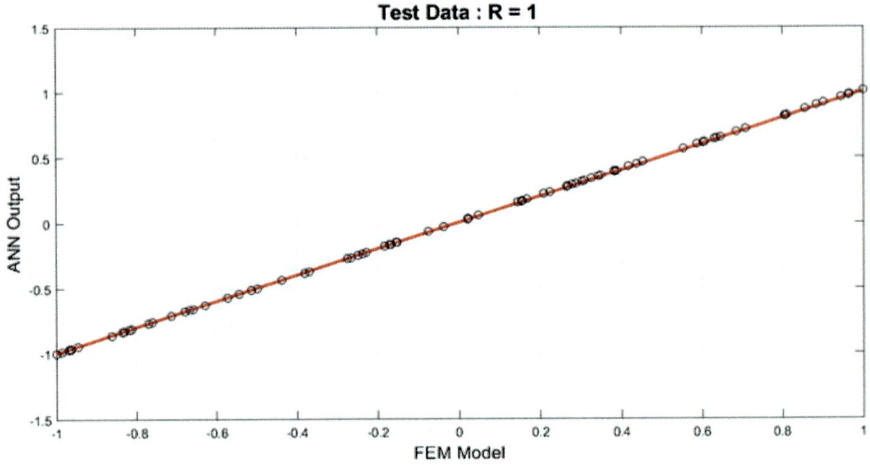

Fig. 18 Linear regression representation relationship among real test data and test data predicted by ANFIS

5 Conclusions

In this paper, a novel method is proposed to detect severities of single damages in RLCPs using an adaptive neuro-fuzzy inference system (ANFIS). and the finite element method (FEM). A nine-node higher-order rectangular element with 5-DoFs is used to write the motion-based governing equation based on FEM. The FEM is applied to compare modal characteristics of the intact and single-damaged RLCP.

Results indicate that the natural frequency of single-damaged RLCP is less than the intact RLCP. These results are consistent with the results reported in the literature and indicate our results' accuracy. Also, the current paper's findings show that the single-damage causes a shift in mode shapes and increases instability. Also, the new model proposed in this study shows that the real training data versus train data predicted by the ANFIS are perfect consistent since R-value is approximately 1. Our findings also indicate that the real test data versus test data predicted by the ANFIS are perfect consistent since R-value is approximately 1. According to results, the MLP model can identify the severity of every single damage scenario in RLCPs with a high precision ($R > 0.9$). Also, results obtained demonstrate that ANFIS can identify and predict the severity of every single damage scenario in RLCPs better than MLP with perfect precision ($R = 0.99999$).

References

1. Khatir S, Tiachacht S, Le Thanh C, Khatir T, Capozucca R, Wahab MA (2020) Damage detection in laminated composite plates based on local frequency change ratio indicator. In: Proceedings of the 13th international conference on damage assessment of structures. Springer, Singapore, pp 887–898
2. Abedi M, Jafari-Talookolaei RA, Valvo PS (2016) A new solution method for free vibration analysis of rectangular laminated composite plates with general stacking sequences and edge restraints. Comput Struct 175:144–156
3. Jafari-Talookolaei RA (2015) Analytical solution for the free vibration characteristics of the rotating composite beams with a delamination. Aerosp Sci Technol 45:346–358
4. Yang JY, Xia BH, Chen Z, Li TL, Liu R (2020) Vibration-based structural damage identification: a review. Int J Robot Autom 35(2)
5. Maity D, Tripathy RR (2005) Damage assessment of structures from changes in natural frequencies using genetic algorithm. Struct Eng Mech 19(1):21–42
6. Dinh-Cong D, Vo-Duy T, Ho-Huu V, Nguyen-Thoi T (2019) Damage assessment in plate-like structures using a two-stage method based on modal strain energy change and Jaya algorithm. Inverse Problems Sci Eng 27(2):166–189
7. Gomes GF, da Cunha SS, Ancelotti AC (2019) A sunflower optimization (SFO) algorithm applied to damage identification on laminated composite plates. Eng Comput 35(2):619–626
8. Barman SK, Maiti DK, Maity D (2020) Damage detection of truss employing swarm-based optimization techniques: a comparison. In: Advanced Engineering Optimization Through Intelligent Techniques. Springer, Singapore, pp 21–37
9. Gerist S, Maheri MR (2019) Structural damage detection using imperialist competitive algorithm and damage function. Appl Soft Comput 77:1–23
10. Liang RY, Hu J, Choy F (1992) Theoretical study of crack-induced eigenfrequency changes on beam structures. J Eng Mech 118(2):384–396
11. Nasiri MR, Mahjoob MJ, Aghakasiri A (2011) Damage detection in a composite plate using modal analysis and artificial intelligence. Appl Compos Mater 18(6):513–520
12. Yam LH, Yan YJ, Jiang JS (2003) Vibration-based damage detection for composite structures using wavelet transform and neural network identification. Compos Struct 60(4):403–412
13. Khatir S, Tiachacht S, Thanh C-L et al (2019) Damage assessment in composite laminates using ANN-PSO-IGA and Cornwell indicator. Compos Struct 230:111509
14. Yu L, Cheng L, Yam LH, Yan YJ, Jiang JS (2007) Online damage detection for laminated composite shells partially filled with fluid. Compos Struct 80(3):334–342

15. Nanda J, Das LD, Choudhury S, Parhi DR (2020) Revelence of multiple breathing cracks on fixed shaft using ANFIS and ANN. In: Innovative product design and intelligent manufacturing systems. Springer, Singapore, pp 599–618
16. Hamidian D, Salajegheh J, Salajegheh E (2018) Damage detection of irregular plates and regular dams by wavelet transform combined adoptive neuro fuzzy inference system. Civil Engineering Journal 4(2):305–319
17. Reddy JN (2003) Mechanics of laminated composite plates and shells: theory and analysis. CRC Press
18. Zainal-Mokhtar K, Mohamad-Saleh J (2013) An oil fraction neural sensor developed using electrical capacitance tomography sensor data. Sensors 13(9):11385–11406
19. Khatir S, Boutchicha D, Le Thanh C, Tran-Ngoc H, Nguyen TN, Abdel-Wahab M (2020) Improved ANN technique combined with Jaya algorithm for crack identification in plates using XIGA and experimental analysis. Theor Appl Fract Mech 102554
20. Tran-Ngoc H, Khatir S, De Roeck G, Bui-Tien T, Wahab MA (2019) An effi-cient artificial neural network for damage detection in bridges and beam-like structures by improving training parameters using cuckoo search algorithm. Eng Struct 199:109637

Damage Identification in Frame Structure Based on Inverse Analysis

Samir Khatir, Samir Tiachacht, Brahim Benaissa, Cuong Le Thanh, Roberto Capozucca, and Magd Abdel Wahab

Abstract Structural damage is a challenging issue in civil and mechanical engineering. Researchers developed several approaches to address it thought out the last few decades, which are also known as large development of metaheuristic algorithms. Their ability to overcome local minima very well compared to local search algorithms helped the raise of damage identification approaches based on the inverse problems. Allowing us to tackle more complex structures and damage parameters like the location and severity. This paper investigates the performance of the of Whale Optimization Algorithm (WOA) technique for model updating and damage identification, by solving an inverse problem to calibrate a Finite Element model of frame structure based on measurements. The performance of WOA is compared to the Salp Swarm Algorithm (SSA) and Particle Swarm Optimization (PSO) from the literature. This study considers an objective function that is based on vibrational frequency, comparing the experimentaly measured and calculated frequencies. To challenge the effectiveness of this application, two cases based on single and multiple damages are presented. The results show that better accuracy is obtained using WOA than Salp Swarm Algorithm (SSA) and Particle Swarm Optimization (PSO). The obtained results clearly indicate that the proposed technique can be used to estimate accurately and efficiently both damage location and severity in frame structures.

S. Khatir (✉) · M. Abdel Wahab
Soete Laboratory, Faculty of Engineering and Architecture, Ghent University, Technologiepark Zwijnaarde 903, B-9052 Zwijnaarde, Belgium
e-mail: samir.khatir@ugent.be

S. Tiachacht
Laboratory of Mechanics, Structure and Energetics (LMSE), Mouloud Mammeri University of Tizi-Ouzou, B.P.N 17 RP, 15000 Tizi-Ouzou, Algeria

B. Benaissa
Department of Mechanical Systems Engineering, Design Engineering Lab, Toyota Technological Institute, Tempaku Ward, Hisakata, 2 Chome−12−1, Nagoya 468-8511 Aichi, Japan

C. Le Thanh
Ho Chi Minh City Open University, Ho Chi Minh City, Vietnam

R. Capozucca
Università Politecnica delle Marche, Ancona, Italy

M. Abdel Wahab (ed.), *Proceedings of the 2nd International Conference on Structural Damage Modelling and Assessment*, Lecture Notes in Civil Engineering 204,
https://doi.org/10.1007/978-981-16-7216-3_15

Keywords Inverse problem · WOA · Model updating · Convergence study · Damage detection

1 Introduction

Structural damage identification consists of three separate stages to define the presence, location, and quantification of the damaged elements. Modified Cornwell Indicator (MCI) was developed by Tiachacht et al. [1] for damage identification in complex structures 3D. The presented results showed that MCI can predict the presence, location, and severity of damaged elements compared with CI. Next, the authors used MCI as an objective function to quantify the damaged elements. Khatir et al. [2, 3] provided the inverse analysis using Particle Swarm Optimization (PSO) and Genetic Algorithm (GA) for damage detection, localization, and quantification in beam composite graphite-epoxy. Ghannadi et al. [4] presented an inverse problem of damage detection using an artificial neural network. The sensitivity analysis carried out for each example showing the performance of an artificial neural network after mode shape expansion was efficiently improved.

Seventekidis et al., [5] suggested an approach based on Convolutional Neural Network (CNN) for multiple damage identification in a composite truss structure. The CNN classifier was trained on the FE generated images combined with a hierarchical multiple damage identification and location scheme. Lee et al. [6] suggested the use of deep neural networks to classify the patterns in the vibrational response of the undamaged and damaged truss structures. The approach made use of the modal strain energy-based index (MSEBI) to reduce connected neural networks. Single or multiple-structural damages have been effectively identified in the case of a 31-bar planar truss. Mousavi et al. [7] suggested a method based on ensemble empirical mode decomposition with adaptive noise for multiple damages identification in the case of steel truss bridge. Cuong et al. [8] suggested an approach for damage identification in truss and frame structures based on improved machine learning using PSO-SVM. And Tran-Ngoc et al. [9] suggested a method based on a hybrid metaheuristic optimization algorithm for damage detection in laminated composite structures. Hybrid Genetic Algorithm and Cuckoo Search (HGACS), were coupled with ANN in the training phase to guarantees that the network possibly determined the best solution fast and avoids getting stuck in local minima. The obtained results showed superior performance to traditional ANN in terms of quality od the results and in computational time.

Xiong et al. [10] presented an Improved Fruit Fly Optimization Algorithm (IFOA) to solve the damage identification problem in a complicated 255-bar truss structure. The suggested optimization method showed better performance in this problem compared to classical methods. Thanh et al. [11] suggested a hybrid heuristic optimization algorithm PSOGSA for damage detection in a truss structure using ECOMAC and frequency in damage detection. Simulation and experimental results presented the efficiency of the method. Kaveh et al. [12] suggested a new optimization algorithm, Guided Water Strider algorithm (GWS), for structural damage detection to reduce the computational costs, using the modal strain energy (MSE) that localize damage and guide the optimization algorithm.

Kaveh et al. [13] proposed a new strategy named Boundary Strategy (BS). To improve the process of damage detection using metaheuristic algorithms. By gradually neutralizing the effects of structural elements that were healthy in the optimization process. Different examples of structures were considered to show the applicability of the proposed technique. Namely the truss structures, beams, and frames, and tested with well-known optimization algorithms including TLBO, GWO, and MFO.

Model order reduction techniques have been used effectively to solve the problem of damage identification [2, 14, 15], Both in studies of elasticity and vibrations. Such an application collects structural response data in order to build a representative model that can be coupled with an optimization algorithm. The introduction of model order reduction significantly reduces the computational cost while keeping the same level of accuracy. Proper orthogonal decomposition (POD) was coupled with eXtended IsoGeometric Analysis (XIGA), for the problem of Single and multiple cracks identifications in plates [16]. In this study, the inverse problem was solved using Jaya and Cuckoo search algorithms. Liang et al. [17] suggested an approach based on POD for real-time identification of damages in multi-layered composite materials, the reduced model is built on heat conduction and heat radiation data issued by FEM.

Nobahari et al. [18] investigated an efficient technique for multiple damages in simple and complex structures. The new process is named flexibility strain energy-based index (FSEBI). The results showed that the proposed indicator could accurately estimate the location of damage and revealed the potential of damage. A two-step approach to determine the location and level of multiple damages was based on the information fusion technique, and GA was investigated in Ref [19]. Ghannadi et al. [20] used Grey Wolf Optimization (GWO) algorithm for SHM of skeletal structures via expanded mode shapes. The objective function was based on natural frequencies and mode shapes. Two experimental examples were studied using a cantilever beam and a truss tower to validate the proposed approach. Multi-verse Optimizer (MVO) was used to solve the damage identification problem by Ghannadi et al. [21]. Two objective functions were used to solve the inverse problem based on the modal assurance criterion (MAC) and modified total modal assurance criterion (MTMAC). Huang et al. [22] presented an approach that employs an improved whale optimization algorithm (LWOA) for damage identification. In this study, the global structure is decomposed into several substructures then the LWOA algorithm is used to calculate the severity of structural damage. Moezi et al. [23] created an improved CS algorithm for open-edge crack identification in cantilever Euler-Bernoulli beam. Depth and location of crack were considered. Next, the structural response used in this study was the natural frequencies obtained from measurements. The performance of the suggested algorithm was compared to results with well-established optimization algorithms. Livani et al. [24] presented a study of damage identification in an Aluminium plate using an enhanced particle swarm optimization (PSO) with a strategy called active/inactive flaw (AIF). The extended spectral finite element method (XSFEM) was employed to obtain the horizontal boundary displacement at selected points. Two types of damages were considered in this study, namely a crack catheterized by the 2D coordinates of the tips, and a circular flaw catheterized by its radius and the 2D coordinates of its center. Benaissa et al. [15, 25, 26] suggested a method for fast crack identification in steel plates, using the vertical displacement in sensor points. A reduced model was built with snapshot data issues from several simulations, using Proper Orthogonal Decomposition and Radial Basis Functions (POD-RBF). For the inverse problem, the performance of several optimization algorithms was compared. Zenzen et al. [27, 28] suggested inverse identification studies based on vibration responses in composite and truss structures using Bat optimization algorithm and ANN. Tiachacht et al. [29] used Genetic Algorithm (GA) for damage identification in 3D using FEM and in the multiple damages, estimation in truss structures. An improved damage indicator called Modified Cornwell Indicator (MCI) in suggested in [1].

In this paper, a convergence study of WO is analysed for model updating and damage detection. Model reduction of FE frame structure combined with WOA based on the inverse problem is investigated to adapt the natural frequencies based on the measurements. The results show that WOA has good convergence and good prediction compared with PSO and SSA.

2 Whale Optimization Algorithm (WOA)

Whales are creatures of fancy. They are considered the world's largest mammals. Up to 30 m long and 180 t weight, an adult whale can grow. This kind of Whale never sleeps because they have to breathe from the ocean floor. Seven different species of this giant mammal such Killer, Sei, humpback, Minke, finback, right, and blue. Whales are known mainly as predators. Half of the brain simply just sleeps. The remarkable thing about whales is that they are viewed with emotion as highly intelligent creatures. The location of prey can be recognized and encircled by humpback whales. The optimal design location in the search space is not known a priori. WOA algorithm assumes that the target prey is close to the optimum or is the current best candidate solution. Other search agents will then update their positions to the best search agent after the best search agent is identified. The following equations represent this behaviour:

$$\vec{D} = \left| \vec{C} \cdot \overrightarrow{X^*}(t) - \vec{X}(t) \right| \tag{1}$$

$$\vec{X}(t+1) = \overrightarrow{X^*}(t) - \vec{A} \cdot \vec{D} \tag{2}$$

where the latest iteration is defined by t, Coefficient vectors are $\vec{A}$ and $\vec{C}$, $\vec{X}$ position vector, and $\overrightarrow{X^*}$ is a position vector of the best solution obtained so far.

$\vec{A}$ and $\vec{C}$ are be calculated in the following formulations:

$$\vec{A} = 2\vec{a} \cdot \vec{r} - \vec{a} \tag{3}$$

$$\vec{C} = 2 \cdot \vec{r} \tag{4}$$

Then a spiral equation is created to approximate the helix-shaped movement of humpback whales between the location of Whale and prey as presented in the following formulation:

$$\vec{X}(t+1) = \overrightarrow{D'} \cdot e^{bl} \cdot \cos(2\pi l) + \overrightarrow{X^*}(t) \tag{5}$$

where $\overrightarrow{D'} = \left|\overrightarrow{X^*}(t) - \vec{X}(t)\right|$

(X, Y): Whale located.

(X^*, Y^*): Prey located.

Humpback whales swim within a decreasing circle around the prey and along a spiral-shaped direction simultaneously. Furthermore, to model this concurrent behaviour, we expect a 50% chance of choosing between either the shrinking encircling process or the spiral model to adjust the location of whales through optimization. The mathematical model is as follows:

$$\vec{X}(t+1) = \begin{cases} \overrightarrow{X^*}(t) - \vec{A} \cdot \vec{D} & if\ p < 0.5 \\ \overrightarrow{D'} \cdot e^{bl} \cdot \cos(2\pi l) + \overrightarrow{X^*}(t) & if\ p \geq 0.5 \end{cases} \tag{6}$$

where p is a random number in [0, 1]. For more details [30].

The mathematical model is as follows:

$$\vec{D} = \left|\vec{C} \cdot \overrightarrow{X_{rand}} - \vec{X}\right| \tag{7}$$

$$\vec{X}(t+1) = \overrightarrow{X_{rand}}(t) - \vec{A} \cdot \vec{D} \tag{8}$$

where $\overrightarrow{X_{rand}}$ is a random position vector (a random whale) chosen from the current population. Some of the possible positions around a particular solution $\vec{A} > 1$ are presented in Fig. 1.

3 Experimental Analysis

In this section, we discuss the formulation of damage identification in a frame structure. The forward problem is issued by the Finite Element Method and calibrated by an inverse problem to fit the vibrational results issued of the experiment.

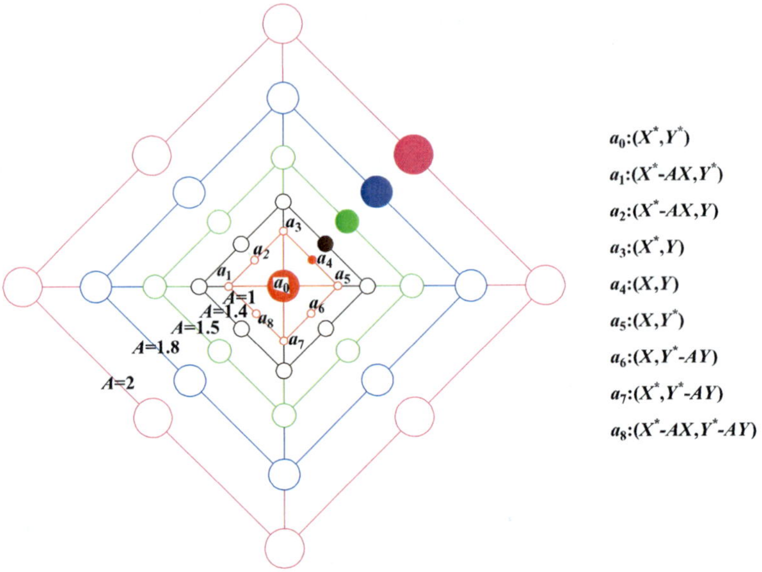

Fig. 1 Exploration mechanism implemented in WOA

A five-story shear frame was analysed on a shaking table to validate FEM using an inverse problem, as shown in Fig. 2. Five accelerometers were provided in this

Fig. 2 Experimental configuration of a five-story shear frame [31]

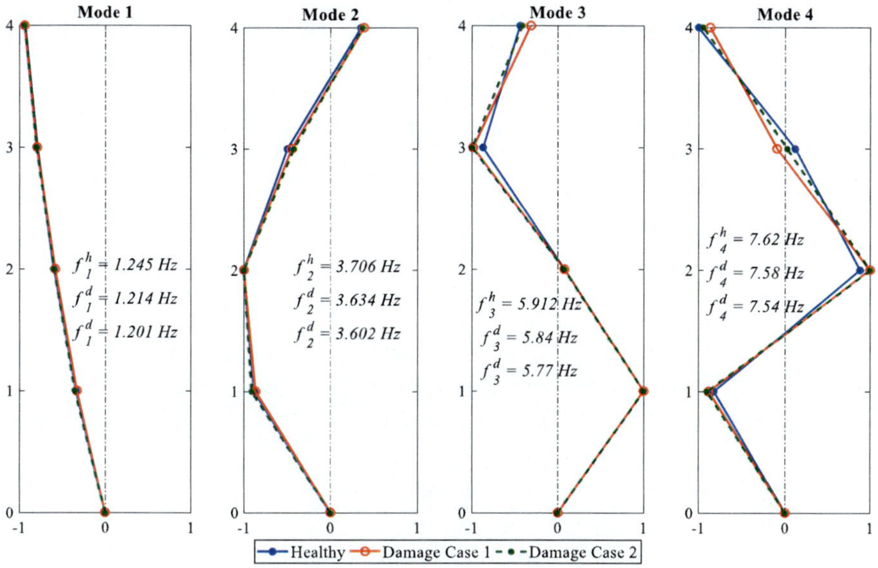

Fig. 3 Experimental mode shapes and natural frequencies of healthy, damage case1 and damage case 2

experiment. The structure was excited by the shaking table for 900 s. This measurement was repeated eight times for each undamaged and damaged case to improve the modal data. The mass and stiffness for each story were 16.09 kg and 11.89 kN/m, respectively. The mode shapes are plotted in Fig. 3.

Calibration of the FEM is performed using the objective function presented in the following formulation:

$$Minimize : fun(x) = \sqrt{\sum_{i=1}^{r} \frac{\left(\omega_i^c - \omega_i^m\right)^2}{\left(\omega_i^m\right)^2}} \qquad i = 1, \ldots, r \tag{9}$$

where ω_i^c is i^{th} frequency computed by optimization technique, ω_i^m is the measured frequency and r denotes the order number of modes used. The optimization algorithm is tasked to minimize this error in order to find the same characteristics that correspond to the measured frequency ω_i^m.

After calibration, the initial FEM stiffness matrix and updated models using WOA are presented in the following matrices:

$$K_{Initial} = 10^4 \times \begin{bmatrix} 2.4104 & -1.1800 & 0 & 0 & 0 \\ -1.1800 & 2.3245 & -1.1445 & 0 & 0 \\ 0 & -1.1445 & 2.3031 & -1.1586 & 0 \\ 0 & 0 & -1.1586 & 2.3680 & -1.2095 \\ 0 & 0 & 0 & -1.2095 & 1.2095 \end{bmatrix} (N/m) \tag{10}$$

$$K_{WOA} = 10^4 \times \begin{bmatrix} 2.3219 & -1.2728 & 0 & 0 & 0 \\ -1.2728 & 2.6920 & -1.4192 & 0 & 0 \\ 0 & -1.4192 & 2.7808 & -1.3616 & 0 \\ 0 & 0 & -1.3616 & 2.7263 & -1.3646 \\ 0 & 0 & 0 & -1.3646 & 1.3646 \end{bmatrix} (N/m) \tag{11}$$

The diagonal mass matrix of the steel frame is as follows:

$$M = \begin{bmatrix} 16.09 & 0 & 0 & 0 & 0 \\ 0 & 16.09 & 0 & 0 & 0 \\ 0 & 0 & 16.09 & 0 & 0 \\ 0 & 0 & 0 & 16.09 & 0 \\ 0 & 0 & 0 & 0 & 16.09 \end{bmatrix} (kg) \tag{12}$$

And the convergence study of the objective function using WOA during 50 iterations is presented in Fig. 4. Table 1 presents calibration results for vibration modes 1 to 4.

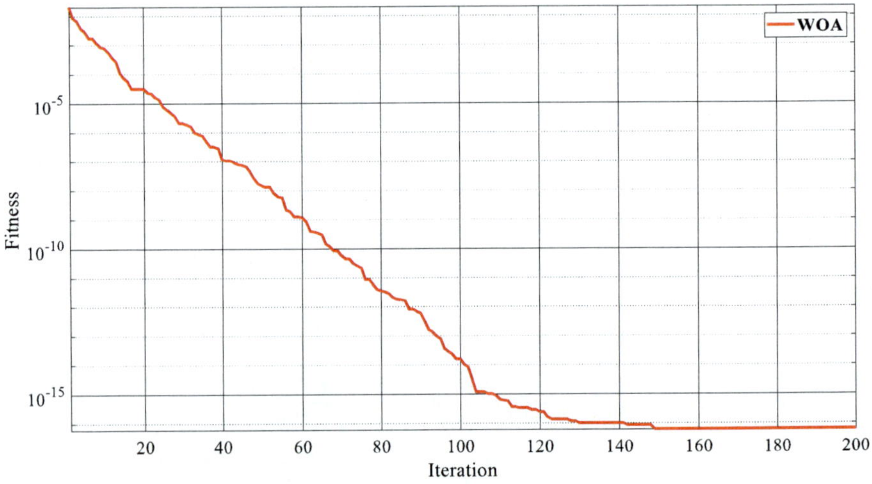

Fig. 4 Evolutionary process using WOA of the FEM updating

Table 1 Natural frequencies (Hz) of five-story shear frame experimental, before and after FEM updating (undamaged and damaged cases)

Mode	Experimental	FEM		Experimental damage	
		Initial	Updated	Case 1	Case2
1	1.245	1.230	1.245	1.214	1.201
2	3.706	3.595	3.706	3.634	3.602
3	5.912	5.666	5.912	5.840	5.770
4	7.620	7.279	7.620	7.580	7.540
1	1.245	1.230	1.245	1.214	1.201

The fitness convergence of WOA, is shown to reach the final position at the 27th iteration. And we are getting very close values after just five iterations.

Table 1 shows that the updated estimations of the truss structure characteristics are equivalent to the experimental results. We then consider that the calibrated FEM model can be used for damage detection.

4 Results and Discussions

In this section, the convergence of damaged cases is plotted in Figs. 5 and 6. The potential of damaged cases with errors are presented in Figs. 7 and 8. In the first case, single damage is presented in the 1st story at 10% level. And in the second case, a double damage scenario is suggested, where the first one is in the 1st story at 10% level, and the second damage in the 3rd story at 10% as well (Tables 2 and 3).

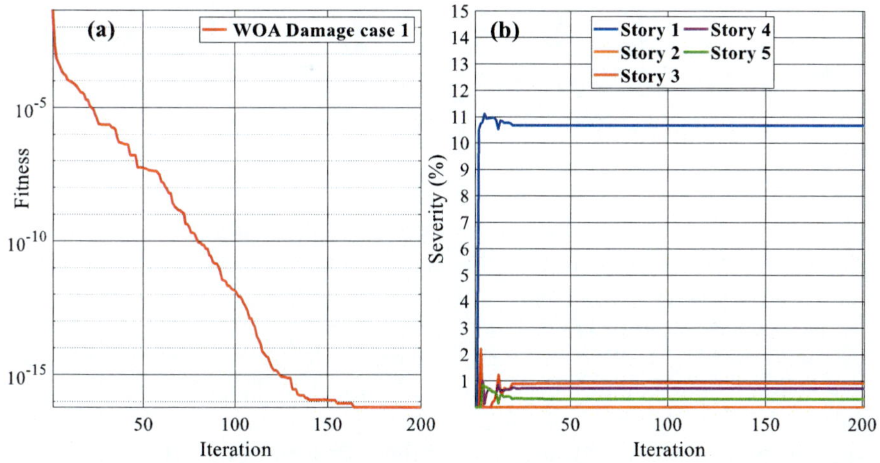

Fig. 5 Evolutionary process using WOA of the damage case 1: **a** Fitness and **b** Damage index

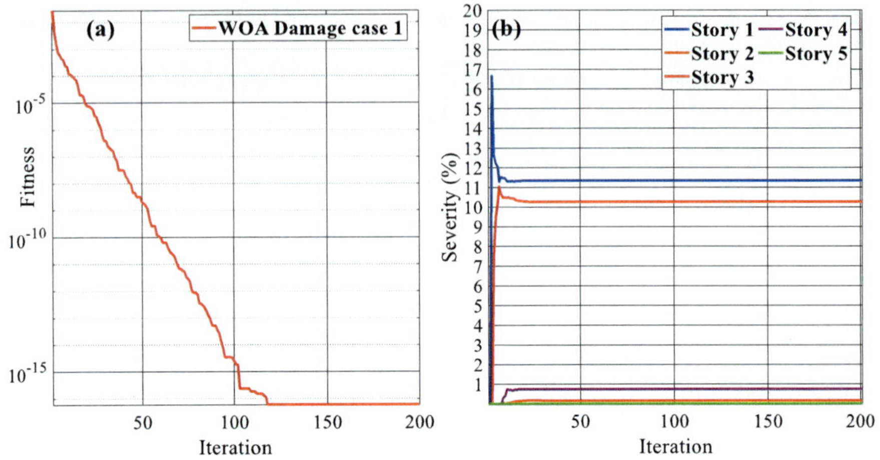

Fig. 6 Evolutionary process using WOA of the damage case 2: **a** Fitness and **b** Damage index

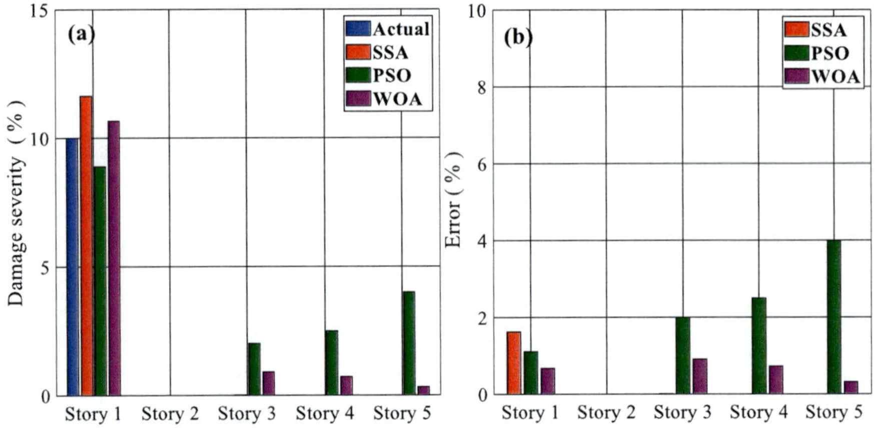

Fig. 7 Results of damage detection for five-story shear frame - Damaged case 1: **a** Damage detection, **b** Error

In the first scenario, the WOA and SSA presented the best performances, while PSO presented the smallest error for the damage at the first story, but it suggested that there are damages in every story with different levels.

In the second scenario, the PSO also suggests an error in every story, but its performance in this case of double damages is not as accurate as in the case of single damage. Similar observations are seen for the SSA algorithm. The error is higher in this case. WOA algorithm, on the other hand, presents the smallest error, but it suggests that there are small damages where there are none. Further results in detail are shown in Tables 4 and 5.

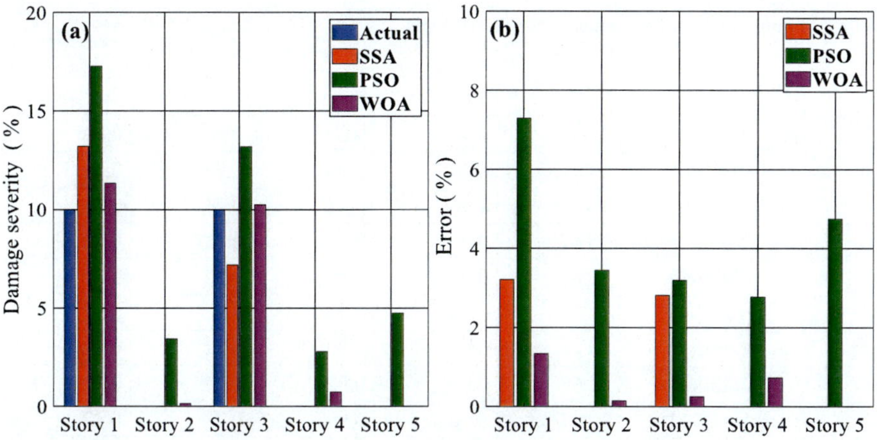

Fig. 8 Results of damage detection for five-story shear frame—Damaged case 2: **a** Damage detection, and **b** Error

The comparison of three optimization techniques for damage localization and quantification of both cases are reported in Tables 4 and 5.

5 Conclusion

This paper tackles the problem of damage identification in a five story—shear frame structure, considering single and multiple damages. The performance of the Whale Optimization Algorithm (WOA) technique is investigated for model updating and damage detection. For the forward problem, the model reduction updated based on stiffness and mass matrixes as well as the updated damage parameters. The objective function based on the measured and calculated frequency. WOA is compared with Salp Swarm Algorithm (SSA) and Particle Swarm Optimization (PSO) from literature. The investigated results have shown that WOA has good convergence compared with other optimization techniques in damage identification and can detect the best solution after a few iterations. This work will be extended for model updating and damage identification in complex structures as future work.

Table 2 Fitness and damage index of the damage case 1

Iteration	Fitness	Severity (%)				
		Story 1	Story 2	Story 3	Story 4	Story 5
1	4.30E-02	0.00	0.00	0.00	0.00	0.00
10	9.33E-05	10.94	0.00	0.25	0.58	0.58
20	1.85E-05	10.67	0.00	0.88	0.74	0.32
30	2.20E-06	10.67	0.00	0.89	0.71	0.32
40	4.00E-07	10.67	0.00	0.90	0.72	0.32
50	5.22E-08	10.67	0.00	0.90	0.72	0.32
60	1.59E-08	10.67	0.00	0.90	0.72	0.32
70	1.38E-09	10.67	0.00	0.90	0.72	0.32
80	8.47E-11	10.67	0.00	0.90	0.72	0.32
90	1.41E-11	10.67	0.00	0.90	0.72	0.32
100	1.37E-12	10.67	0.00	0.90	0.72	0.32
110	7.87E-14	10.67	0.00	0.90	0.72	0.32
120	1.94E-15	10.67	0.00	0.90	0.72	0.32
130	7.50E-16	10.67	0.00	0.90	0.72	0.32
140	1.14E-16	10.67	0.00	0.90	0.72	0.32
150	1.10E-16	10.67	0.00	0.90	0.72	0.32
160	8.47E-17	10.67	0.00	0.90	0.72	0.32
170	6.11E-17	10.67	0.00	0.90	0.72	0.32
180	6.11E-17	10.67	0.00	0.90	0.72	0.32
190	6.11E-17	10.67	0.00	0.90	0.72	0.32
200	6.11E-17	10.67	0.00	0.90	0.72	0.32
Actual Damage		10%	0%	0%	0%	0%

Table 3 Fitness and damage index of the damage case 2

Iteration	Fitness	Severity (%)				
		Story 1	Story 2	Story 3	Story 4	Story 5
1	4.30E-02	0.00	0.00	0.00	0.00	0.00
10	9.33E-05	11.30	0.03	10.47	0.73	0.00
20	1.85E-05	11.33	0.15	10.27	0.73	0.00
30	2.20E-06	11.34	0.15	10.26	0.73	0.00
40	4.00E-07	11.34	0.15	10.26	0.73	0.00
50	5.22E-08	11.34	0.15	10.26	0.73	0.00
60	1.59E-08	11.34	0.15	10.26	0.73	0.00
70	1.38E-09	11.34	0.15	10.26	0.73	0.00
80	8.47E-11	11.34	0.15	10.26	0.73	0.00
90	1.41E-11	11.34	0.15	10.26	0.73	0.00
100	1.37E-12	11.34	0.15	10.26	0.73	0.00
110	7.87E-14	11.34	0.15	10.26	0.73	0.00
120	1.94E-15	11.34	0.15	10.26	0.73	0.00
130	7.50E-16	11.34	0.15	10.26	0.73	0.00
140	1.14E-16	11.34	0.15	10.26	0.73	0.00
150	1.10E-16	11.34	0.15	10.26	0.73	0.00
160	8.47E-17	11.34	0.15	10.26	0.73	0.00
170	6.11E-17	11.34	0.15	10.26	0.73	0.00
180	6.11E-17	11.34	0.15	10.26	0.73	0.00
190	6.11E-17	11.34	0.15	10.26	0.73	0.00
200	6.11E-17	11.34	0.15	10.26	0.73	0.00
Actual Damage		10%	0%	10%	0%	0%

Table 4 Comparison between detected and actual results from the five-story shear frame-Damaged case 1

Story	1	2	3	4	5
Actual	10.00%	0%	0%	0%	0%
SSA [31]	11.63%	0%	0.005%	0%	0%
Error	1.63%	0%	0.005%	0%	0%
PSO [31]	8.90%	0%	2.00%	2.50%	4.00%
Error	1.10%	0%	2.00%	2.50%	4.00%
WOA	10.67%	0.00%	0.90%	1%	0%
Error	0.67%	0.00%	0.90%	1%	0%

Table 5 Comparison between detected and actual results from five-story shear frame-Damaged case 2

Story	1	2	3	4	5
Actual	10.00%	0%	10.00%	0%	0%
SSA [31]	13.22%	0%	7.18%	0%	0%
Error	3.22%	0%	2.82%	0%	0%
PSO [31]	17.30%	3.45%	13.20%	2.78%	4.75%
Error	7.30%	3.45%	3.20%	2.78%	4.75%
WOA	11.34%	0.15%	10.26%	0.73%	0.00%
Error	1.34%	0.15%	0.26%	0.73%	0.00%

Acknowledgements The first author acknowledges the funding of the postdoctoral fellowship BOF20/PDO/045 provided by Bijzonder Onderzoeksfonds (BOF), Ghent University.

References

1. Tiachacht S et al (2018) Damage assessment in structures using combination of a modified Cornwell indicator and genetic algorithm. Eng Struct 177:421–430
2. Khatir S et al (2017) Multiple damage detection in composite beams using particle swarm optimization and genetic algorithm. Mechanics 23(4):514–521
3. Khatir S et al (2015) Damage detection and localization in composite beam structures based on vibration analysis. Mechanics 21(6):472–479
4. Ghannadi P, Kourehli SS (2021) An effective method for damage assessment based on limited measured locations in skeletal structures. Adv Struct Eng 24(1):183–195
5. Seventekidis P, Giagopoulos D (2021) A combined finite element and hierarchical Deep learning approach for structural health monitoring: test on a pin-joint composite truss structure. Mech Syst Sig Proc 157:107735
6. Lee S et al (2021) Damage quantification in truss structures by limited sensor-based surrogate model. Appl Acoust 172: 107547
7. Mousavi AA et al (2021) Structural damage detection method based on the complete ensemble empirical mode decomposition with adaptive noise: a model steel truss bridge case study. Struct Health Monit 2021:14759217211013535
8. Cuong-Le T et al (2021) An efficient approach for damage identification based on improved machine learning using PSO-SVM. Eng Comput 2021:1–16
9. Tran-Ngoc H et al (2021) Efficient artificial neural networks based on a hybrid metaheuristic optimization algorithm for damage detection in laminated composite structures. Compos Struct 262:113339
10. Xiong C, Lian S (2021) Structural damage identification based on improved fruit fly optimization algorithm. KSCE J Civ Eng 25(3):985–1007
11. Thanh B-T, Nguyen QT, Ho LV (2021) A hybrid heuristic optimization algorithm PSOGSA coupled with a hybrid objective function using ECOMAC and frequency in damage detection. J Mater Eng Struct JMES 8(1):31–45
12. Kaveh A, Rahmani P, Eslamlou AD (2021) Guided water strider algorithm for structural damage detection using incomplete modal data. Iran J Sci Technol, Trans Civ Eng 2021:1–18
13. Kaveh A, Hosseini SM, Zaerreza A (2021) Boundary strategy for optimization-based structural damage detection problem using Metaheuristic algorithms. Period Polytech Civ Eng 65(1):150–167

14. Benaissa B et al (2016) Crack identification using model reduction based on proper orthogonal decomposition coupled with radial basis functions. Struct Multidiscip Optim 54(2):265–274
15. Benaissa B et al (2017) Application of proper orthogonal decomposition and radial basis functions for crack size estimation using particle swarm optimization. In: Journal of Physics: Conference Series. IOP Publishing
16. Khatir S, Wahab MA (2019) Fast simulations for solving fracture mechanics inverse problems using POD-RBF XIGA and Jaya algorithm. Eng Fract Mech 205:285–300
17. Liang Y et al (2021) A reduced-order modelling for real-time identification of damages in multi-layered composite materials. Inverse Probl Sci Eng 29(1):73–94
18. Nobahari M, Seyedpoor SM (2013) An efficient method for structural damage localization based on the concepts of flexibility matrix and strain energy of a structure. Struct Eng Mech: Int J 46(2):231–244
19. Guo HY, Li ZL (2009) A two-stage method to identify structural damage sites and extents by using evidence theory and micro-search genetic algorithm. Mech Syst Signal Process 23(3):769–782
20. Ghannadi P et al (2020) Efficiency of grey wolf optimization algorithm for damage detection of skeletal structures via expanded mode shapes. Adv Struct Eng 23(13):2850–2865
21. Ghannadi P, Kourehli SS (2020) Multiverse optimizer for structural damage detection: numerical study and experimental validation. Struct Des Tall Spec Build 29(13):e1777
22. Huang M, Cheng X, Lei Y (2021) Structural damage identification based on substructure method and improved whale optimization algorithm. J Civ Struct Heal Monit 11(2):351–380
23. Moezi SA, Zakeri E, Zare A (2018) A generally modified cuckoo optimization algorithm for crack detection in cantilever Euler-Bernoulli beams. Precis Eng 52:227–241
24. Livani M, Khaji N, Zakian P (2018) Identification of multiple flaws in 2D structures using dynamic extended spectral finite element method with a universally enhanced meta-heuristic optimizer. Struct Multidiscip Optim 57(2):605–623
25. Benaissa B et al (2016) Crack identification using model reduction based on proper orthogonal decomposition coupled with radial basis functions. Struct Multidiscip Optim 2016:1–10
26. Samir K et al (2018) Damage detection in CFRP composite beams based on vibration analysis using proper orthogonal decomposition method with radial basis functions and cuckoo search algorithm. Compos Struct 187:344–353
27. Zenzen R et al (2018) A damage identification technique for beam-like and truss structures based on FRF and Bat algorithm. Comptes Rendus Mécanique 346(12):1253–1266
28. Zenzen R et al (2020) A modified transmissibility indicator and artificial neural network for damage identification and quantification in laminated composite structures. Compos Struct 248:112497
29. Tiachacht S et al (2018) Structural health monitoring of 3D frame structures using finite element modal analysis and genetic algorithm. J Vibroeng 20(1):202–214
30. Mirjalili S, Lewis A (2016) The whale optimization algorithm. Adv Eng Softw 95:51–67
31. Ghannadi P, Kourehli SS (2019) Model updating and damage detection in multi-story shear frames using Salp Swarm Algorithm. Earthq Struct 17(1):63–73